Sustainable
Sludge
Management
Resource Recovery for
Construction Applications

Sustainable
Sludge
Management

Resource Recovery for
Construction Applications

Kuan-Yeow Show
Hohai University, China

Joo-Hwa Tay
University of Calgary, Canada

Duu-Jong Lee
National Taiwan University, Taiwan

World Scientific

NEW JERSEY · LONDON · SINGAPORE · BEIJING · SHANGHAI · HONG KONG · TAIPEI · CHENNAI · TOKYO

Published by

World Scientific Publishing Co. Pte. Ltd.

5 Toh Tuck Link, Singapore 596224

USA office: 27 Warren Street, Suite 401-402, Hackensack, NJ 07601

UK office: 57 Shelton Street, Covent Garden, London WC2H 9HE

Library of Congress Cataloging-in-Publication Data

Names: Show, Kuan-Yeow, author. | Tay, Joo Hwa, author. | Lee, D. J. (Duu-Jong), author.

Title: Sustainable sludge management : resource recovery for construction applications / by
Kuan-Yeow Show (Hohai University, China), Joo-Hwa Tay (University of Calgary, Canada),
Duu-Jong Lee (National Taiwan University, Taiwan).

Description: [Hackensack] New Jersey : World Scientific, 2018. |
Includes bibliographical references and index.

Identifiers: LCCN 2018023988 | ISBN 9789813238251 (hc : alk. paper)

Subjects: LCSH: Sewage sludge precipitants--Recycling. | Sewage sludge--Recycling. |
Sewage--Purification--By-products. | Building materials.

Classification: LCC TD772 .S56 2018 | DDC 628.3/8--dc23

LC record available at https://lccn.loc.gov/2018023988

British Library Cataloguing-in-Publication Data

A catalogue record for this book is available from the British Library.

For any available supplementary material, please visit
https://www.worldscientific.com/worldscibooks/10.1142/10938#t=suppl

Desk Editor: Herbert Moses

Typeset by Stallion Press
Email: enquiries@stallionpress.com

Printed in Singapore

About the Authors

 Dr. **Kuan-Yeow Show** graduated with a B.Eng. (Civil) degree from the National University of Singapore in 1988. He obtained his Master and Ph.D. degrees in Environmental Engineering in 1993 and 1996, respectively, from Nanyang Technological University, Singapore. He had served as Tenured Associate Professor in Nanyang Technological University, and as Professor and Dean of the Faculty of Engineering & Green Technology cum SP Setia Endowed Chair Professor of Environmental Engineering & Green Technology, University of Tunku Abdul Rahman (UTAR), Malaysia. He is currently serving as the Director, Environmental Technology Institute, Puritek Co. Ltd., China and Zhejiang Juneng Co., Ltd, China. He is also currently serving as Visiting Professor, College of Environment, Hohai University, Nanjing, China.

Dr. Show's areas of specialization include biogranulation technology and high-performance anaerobic systems for wastewater treatment. He has been awarded several patents including United States Patent, International Patent, European Patent, Singapore Patent, Hong Kong Patent, Korea Patent and China Patent on a biogranulation technology for wastewater treatment. He has received several professional awards including the prestigious *Singapore National Technology Award 2003* for "Biogranulation Technologies for High-Performance Biological Wastewater

Treatment". The award recognizes scientists and engineers in Singapore who have made outstanding contributions to R&D resulting in significant technology with industrial applications. He also participated in the research team receiving the *Engineering Achievement Awards 2004* given by the Singapore Institution of Engineers and *The Enterprise Challenge Innovator Award* in 2001 by the Singapore Prime Minister's Office for innovation on biogranulation technology.

Arising from numerous successful applications of the biogranulation technologies for treatment of recalcitrant and toxic industrial effluents from petrochemicals, coal chemicals, specialty chemicals, textile and dyeing, and centralized WWTPs in industrial parks, he received the 2015 *West Lake Friendship Award* — the highest honor for foreign experts who have made significant contributions to the economic development in China (2015 浙江省外国专家西湖友谊奖), *2014 National 1000 Talents Program, China* (千人计划国家特聘专家), *2014 Zhejiang Province 1000 Talents Program* (浙江省千人计划国家特聘专家), and *2012 Shanghai 1000 Talents Program* (上海千人计划特聘专家). On 5 Feb 2016, he was invited for Foreign Experts Recommendation Symposium cum Lunar New Year Dinner with the Premier Li Keqiang held at the Great Hall of People, Beijing (李克强总理人民大会堂春节座谈会及共进晚餐).

He has published 152 technical papers in refereed international journals and conferences, 86 research reports, forums, short courses etc., 2 editorships for book and journal issue, 1 co-authored book and 30 book chapters. He has been awarded/filing 7 patents including 1 United States Patent 6793822, 1 International Patent WO 2003/070649, 1 European Patent EP14764031, 1 Singapore Patent 2004017703, 1 Hong Kong Patent 1075880, 1 Korea Patent 1020047004869 and 1 China Patent CN 03801641.9 on a biogranulation technology for wastewater treatment. His research outputs are documented in a number of renowned international journals with high research impact which include the *Biotechnology Advances* (Impact Factor 10.597), *Water Research* (Impact Factor 6.942), *Bioresource Technology* (Impact Factor 5.651), *Biotechnology & Bioengineering* (Impact Factor 4.481), *Applied Microbiology and Biotechnology* (Impact Factor 3.420) and *International Journal of Hydrogen Energy* (Impact Factor 3.582). His publications have been widely cited in many renowned international journals, conference proceedings, books and chapters. A citation search by Google Scholar database on Dr. Show's research impact and citations indicated a total citation count over 5000 and an author h-index of 36.

Dr. Joo-Hwa Tay is a Professor in Civil Engineering Department of the University of Calgary, Canada, and CRC (Canada Research Chairs) Tier 1 Chair in Sustainable Water Engineering. Prior to joining the University of Calgary, he was a Professor at Nanyang Technological University, Singapore for 30 years. Professor Tay graduated from the National Taiwan University, Taiwan in 1971 with a BSc degree in Civil Engineering. He obtained an MSc degree in Hydraulic and Water Resources Engineering from University of Cincinnati, USA in 1972, and a Ph.D. in Environmental Engineering from University of Toronto, Canada in 1976. He is a registered PEng in both Singapore and Ontario, Canada. He has lectured at universities in Canada, United States and Singapore since 1976.

He has received four Singapore National Awards for his outstanding research and inventions. He has filed twelve patents since 2002. He is the Editor of the *Journal of Hazardous Materials*. He was also an Associate Editor for the *ASCE Journal of Environmental Engineering*. His current research interests focus on the biotechnological application on water and wastewater treatment, sludge, hazardous and industrial waste management, waste recycling and reuse.

Professor Tay has developed various processes to convert wastes into non-conventional construction materials. Some of these processes have been commercialized. He has written more than 35 books/book chapters/monographs and published more than 660 technical papers on environmental engineering and management including 360 SCI Journal papers, with Citation Index of >10,000, and h-Index of 53.

Professor Tay is a member of the Singapore Ministry of the Environment's Advisory Committee on Hazardous Substances & Toxic Wastes since 1991. He was also a member of the Resource Panel of Singapore Government Parliamentary Committee. The Committee comprised of Members of Parliament and Experts to provide advice to the Minister of the Environment and Water Resources on various water and environment issues. He was the Head of the Division of Environmental and Water Resources Engineering in Nanyang Technological University from 1991 to 2005, and Director and CEO of the Singapore National Institute of Environmental Science & Engineering from 2001 to 2008.

Professor Tay has served as consultant to the World Bank, Asian Development Bank, United Nations organizations (e.g. UNEP, UNDP, UNIDO, UNESCAP, UNCRD), World Health Organization (WHO), and several other International Technical Assistant Programs on water and wastewater treatment, environmental engineering and management in the Asia-Pacific region for the last 20 years. He also provides consultancy services on environmental management issues to the governments and industries in the Asia-Pacific region.

In the last 10 years, Professor Tay has designed more than 30 treatment processes for the water and wastewater management in the Asia-Pacific region. He has provided consulting services to Asian Productivity Organization on more than 10 demonstration projects on cleaner production in Asian countries. He has also offered more than 100 training courses for senior staff on water and wastewater treatment, environmental management, industrial and hazardous wastes management, ISO 14000 — Environmental Management System, Environmental Impact Assessment (EIA), Green Productivity, Waste Minimization and Cleaner Production Programmes in more than 20 Asian countries.

Dr. Duu-Jong Lee received his B.Eng. from National Taiwan University, Taiwan in 1984 and his Ph.D. from the same university in 1989.

Prior to joining National Taiwan University in 1992, Dr. Lee was serving as Associate Professor in the Department of Chemical Engineering, Yuan-Ze Institute of Technology, Taiwan. He was appointed as Director of the YZIT Environmental Engineering Research Centre in 1991. After joining

NTU, Dr. Lee was serving as Associate Professor at Department of Chemical Engineering, and was promoted as full Professor in 1996.

Dr. Lee's expertise includes bioenergy recovery from waste, drinking water production using membrane, sludge treatment, and microscale heat and mass transport processes. He has served as an investigator for a number of industrial R&D projects funded by industrial collaborators and government agencies, and has been serving as a consultant to the industry in Taiwan in the area of wastewater treatment. He has engaged in editorial works for *Bioresource Technology, Chemical Engineering Science, Drying Technology, Separation Science & Technology, Applied Energy, Advanced Powder Technology* and others.

He has worked on Biohydrogen production from biomass, Carbon Credits from high-performance anaerobic systems, Accelerated startup and operation of anaerobic reactors, Microbial granulation in wastewater treatment, Ultrasound applications in sludge and wastewater treatment, and Conversion of sludge and wastes into engineering materials.

He has also received numerous awards such as National Science Council Outstanding Research Award (Taiwan MOST), National Science Council Specially Appointed Researcher Award (Taiwan MOST), Outstanding Fellow Award (Taiwan MOST), Academic Award (Taiwan MoE), and National Chair Award (Taiwan MoE).

Dr. Lee has published about 810 technical papers in SCI international journals and about 450 papers in international and local conferences. He has one authored book, one co-authored book and seven book chapters. He has been awarded one Singapore Patent and eight ROC patents on biogranulation technology and wastewater treatment.

A citation record search dated 2 December 2017 on Dr. Lee's journal publications based on ISI WoS indicated a total citation count of 16050 and an H-Index score of 60. Dr. Lee has six papers that have been listed on Essential Science Indicators^{SM} as Highly Cited Papers.

Contents

List of Figures

List of Tables

1

Introduction

Converting sludge into construction materials is deemed to be a sustainable approach to alleviating sludge disposal problems and conserving natural resources. The problems of disposal and depleting raw material resources will be drastically reduced if sludge can be put to large-scale economic applications. This review collates the state-of-the-art development in recycling sludge into major construction materials with in-depth discussion on properties of the products such as bricks, aggregates, and cement replacement materials, their processing technology and technical suitability for construction applications, and leaching assessment of toxic chemicals from the products. Challenges and prospects of sludge recycling are also outlined.

The quantities of sludge produced from municipal wastewater treatment are ever-increasing, arising from industrial growth, construction of new treatment plants and upgrading of existing facilities. A study indicated that for the European market alone, demand for sludge treatment is growing at an annual rate of 6.2% from US$1.95 billion to US$2.77 billion by 2010 (Frost and Sullivan, 2004). Conversely, disposal of sludge from wastewater treatment creates increasingly difficult problems for many municipalities in meeting increasingly stringent environmental regulations. In addressing sludge disposal issue, one of the UN directives is to protect water, air and land environments in the application of sludge (Omer, 2008).

Currently most of the sludge is disposed of by landfilling and spreading on land, while incineration only reduces the volume of sludge, and the remaining residues still require ultimate disposal in the landfill. All these disposal options have varying degrees of impact on the environment. With escalating pressure from regulators and the general public, landfilling of sludge is being phased out in many municipalities because of potential water and land pollution caused by leachate and contaminated soil, dwindling land space, and the emission of a potent greenhouse gas, viz. methane (Cherubini *et al.*, 2009; Foley *et al.*, 2009). While the future of landfilling appears to be gloomy, viable and sustainable disposal alternatives are what scientists and engineers have long sought after.

In the EU, targets were set relative to 2000 levels, that reductions in sludge volume be set to 20% by 2010 and further to half by 2050, whereby the achievement of the objectives are mainly through (a) prevention of waste, (b) waste recovery through reuse, recycle and energy

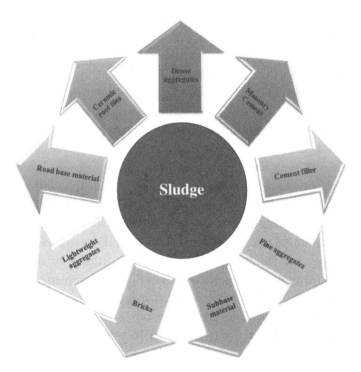

Figure 1.1.   Construction products made of sludge

recovery, (c) improved treatment conditions and (d) regulation of transport (Fytili and Zabaniotou, 2008). Sludge contains matter and energy that can be reused beneficially. Extensive global effort has been made in developing new technologies to convert sludge into useful products, and studies have been carried out to explore possible applications of sludge and sludge ash as construction materials. In addition to alleviating disposal problems as mentioned, recycling of wastewater sludge as construction materials would bring about economic, ecological and energy-saving advantages. This chapter reviews the state-of-the-art development in reusing sludge as some major construction materials. A research-based development of construction products made of sludge large-scale applications, namely bricks, aggregates, cement replacement materials, road base materials, ceramic roofing tiles, and so forth (Figure 1.1), is outlined. In-depth discussions on properties of the products, the production/processing technology, and their technical suitability for construction applications are also presented. Challenges including leaching potential of heavy metals or toxic chemicals from the products and prospect of sludge recycling are also outlined.

2

Characteristics of Sludge and Sludge Ash

Sludge is an odious, semisolid residual that resembles thick soft mud produced from the solid–liquid separation processes in wastewater treatment. It is usually very inconsistent in its composition and most often unmanageable. Contaminants removed from the wastewater are concentrated in the sludge, which pose serious problem to the health and environment if it is not properly managed. On the other hand, sludge could become a valuable and renewable resource with proper treatment and prudent management. Sludge usually undergoes a series of treatment steps involving thickening, stabilization, conditioning, dewatering, and biological or chemical degradation prior to final disposal. Figure 2.1 outlines typical process alternatives and flow for sludge processing and final disposal or reuse.

The characteristics of sludge produced in wastewater treatment vary invariably from plant to plant, due to the great difference in wastewater origin, and in the designed and operation of wastewater treatment plants. It is essential to analyze the levels of contaminants and constituents in the sludge as that would help establish the options for engineering applications or feasible reutilization of the sludge. Typical data on chemical composition of untreated and digested sludges are presented in Table 2.1. Data on the physical characteristics and quantities of sludge produced from various treatment processes and operations are listed in Table 2.2.

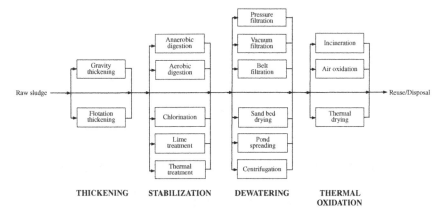

THICKENING   STABILIZATION   DEWATERING   THERMAL
OXIDATION

Figure 2.1.   Typical process alternatives and flow for sludge processing and final disposal

Table 2.1.   Chemical properties of primary sludge (adapted from Metcalf and Eddy, 2004)

Property	Unit	Primary sludge	Digested primary sludge
Dry total solids (TS)	(%)	2–8	6–12
Volatile solids	(% TS)	60–80	30–60
Grease and fats	(% TS)	—	—
Ether soluble	—	6–30	5–20
Ether extract	—	7–35	—
Protein	(% TS)	20–30	15–20
Nitrogen	(% TS)	1.5–4.0	1.6–6.0
Phosphorus as P_2O_5	(% TS)	0.8–2.8	1.5–4.0
Potash as K_2O	(% TS)	0.0–1.0	0.0–3.0
Cellulose	(% TS)	8–15	8–15
Iron (not as sulfide)	(% TS)	2–4	3–8
Silica	(% TS)	15–20	10–20
pH	—	5.0–8.0	6.5–7.5
Alkalinity	mg/L as $CaCO_3$	500–1,500	2,500–3,500
Organic acids	mg/L as HAc	200–2,000	100–600
Energy content	BTU/lb	10,000–12,500	4,000–6,000

Table 2.2. Physical properties and quantities of sludge produced from various wastewater treatment operations and processes (adapted from Metcalf and Eddy, 2004)

Treatment operation or process	Specific gravity of sludge solids	Specific gravity of sludge	Dry solids (kg/1,000 m³)
Primary sedimentation	1.4	1.02	110–170
Activated sludge (waste sludge)	1.25	1.005	70–100
Trickling filter (waste sludge)	1.45	1.025	55–90
Extended aeration (waste sludge)	1.30	1.015	80–120
Aerated lagoon (waste sludge)	1.30	1.01	80–120
Filtration	1.20	1.005	10.20
Algae removal	1.20	1.005	10–20
Lime addition to primary clarifiers for phosphorus removal			
Low lime (350–500 mg/L)	1.9	1.04	250–400
High lime (800–1,600 mg/L)	2.2	1.05	600–1,280
Suspended-growth denitrification	1.20	1.005	—
Roughing filters	1.28	1.02	—

In the laboratory, sludge samples are normally dried in an oven at a temperature of 105°C and tested for various properties such as specific gravity (SG), pH, and loss on ignition. Typical characteristics of municipal and industrial sludge are listed in Table 2.3. The municipal sludge and industrial sludge have moisture content levels ranging from 50.5% to 70.0% and 32.2% to 50.7%, respectively. The SG of municipal sludge ranged from 1.64 to 1.72 while the industrial sludge has higher SG values ranging from 2.67 to 2.70, due to its inorganic nature. The municipal sludge was found to have pH levels in the range of 8.1–8.5, which is slightly alkaline, compared to the industrial sludge, which has pH levels in the range of 7.7–7.8. The loss on ignition of the municipal sludge was determined within the range of 59.2–60.8%. The industrial sludge has levels of loss on ignition in the range of 5.97–12.94%, which are lower compared to that of the municipal sludge, indicating that it contains less volatile solids.

Table 2.3.   Properties of sludge

Property	Municipal sludge (Tay, 1987a)		Industrial sludge (Show *et al.*, 2005, 2006a)	
	Range	Average	Range	Average
Specific gravity	1.64–1.72	1.69	2.67–2.70	2.68
pH	8.1–8.5	8.30	7.7–7.8	7.75
Moisture content (%)	50.5–70.0	60.0	32.2–50.7	43.0
Loss-on-ignition (%)	59.2–60.8	60.0	5.97–12.94	9.33

The loss on ignition represents the carbon dioxide, water and any oxidizable elements present in the materials.

Tables 2.4–2.6 and Figure 2.2 show the comparison of results of chemical and physical properties of wastewater sludge ashes obtained from various studies (Gray and Penessis, 1972; Arakawa *et al.*, 1984; Kato and Takesue, 1984; Elkins *et al.*, 1985; Kurth, 1984; Tay, 1987a, 1987b, 1987c; Bhatty and Reid, 1989a, 1989b; Tay and Yip, 1989; Tay and Show, 1992a, 1992b; Tay *et al.*, 2000; Show *et al.*, 2005, 2006a, 2006b). Properties discussed include chemical and mineralogical composition, SG, bulk density, water absorption, particle density, and porosity.

Chemical composition of the sludge ashes obtained from the studies is listed in Tables 2.4 and 2.5. The sludges were incinerated at temperatures ranging from 550°C to 1700°C. The incineration temperature for each ash is given where information is available. Elements shown in Table 2.2 were quantitatively determined as oxides. Of all the sludge ashes tested, the four major oxides present are silica (SiO_2), lime (CaO), alumina (Al_2O_3) and iron oxides (Fe_2O_3), with these oxides falling within the ranges of 14.39–57.67%, 1.75–36.92%, 4.63–16.41%, and 3.44–29.14%, respectively. The sludges were high in silica and alumina because these compounds are common in earth metals. Lime is a minor constituent in treatment plants that use organic polymers as flocculation and sludge dewatering aids. On the other hand, ashes collected from plants which use lime in lieu of polymers exhibit high lime contents. The lime enhances the strength and durability properties of the sludge ash when it is mixed with water and compacted (Gray and Penessis, 1972).

Table 2.4.   Chemical analysis of wastewater sludge ashes (%)

Ash source	SiO_2	Al_2O_3	Fe_2O_3	MgO	CaO	SO_3	Na_2O	K_2O	P_2O_5	Loss on ignition	Firing temperature (°C)	Reference
Pontiac[a]	32.54	9.60	9.47	2.07	36.92	0.01	0.41	0.66	7.01	1.00	1400–1700	Gray and Penessis (1972)
St. Paula[a]	24.87	13.48	10.81	2.61	33.35	2.71	0.26	0.12	9.88	1.62	NA	Gray and Penessis (1972)
Kansas City[b]	57.67	15.00	8.50	0.85	8.64	3.42	0.45	0.35	4.43	0.31	1500–1600	Gray and Penessis (1972)
Saginaw[a]	28.18	4.63	8.68	2.20	29.86	2.87	0.32	0.07	3.86	15.18	1400–1700	Gray and Penessis (1972)
South Tahoe	23.85	16.34	3.44	2.12	29.76	2.79	0.73	0.14	6.87	2.59	1400–1600	Gray and Penessis (1972)
Cromwell	14.39	4.73	24.40	1.35	26.39	1.68	0.13	0.07	8.63	14.67	NA	Gray and Penessis (1972)
Cuyahoga	28.85	10.20	14.37	2.13	26.37	5.04	0.18	0.25	9.22	1.94	NA	Gray and Penessis (1972)

(*Continued*)

Table 2.4.   (*Continued*)

Ash source	SiO_2	Al_2O_3	Fe_2O_3	MgO	CaO	SO_3	Na_2O	K_2O	P_2O_5	Loss on ignition	Firing temperature (°C)	Reference
Bhatty	27.03	14.36	8.22	3.21	20.97	0.84	0.52	0.63	20.20	0.20	NA	Bhatty and Reid (1989a)
Kato[a]	29.86	11.18	11.09	1.86	25.19	2.92	0.81	0.89	5.41	5.60	1050	Kato and Takesue (1984)
Katob	39.55	16.41	11.75	2.46	8.36	0.84	1.84	1.36	10.10	2.67	1050	Kato and Takesue (1984)
Kurth	50.50	12.10	9.50	2.00	11.70	NA	NA	1.70	11.00	NA	1000–1100	Kurth (1984)
Arakawa[a]	22.50	8.50	15.50	3.50	32.50	NA	0.75	0.65	5.50	NA	1100	Arakawa et al. (1984)
Arakawa[b]	43.00	15.00	12.50	1.80	7.00	NA	0.36	1.70	4.80	NA	1100	Arakawa et al. (1984)
Tay and Show[b]	20.33	14.64	20.56	2.07	1.75	7.80	0.51	1.81	NA	10.45	550	Tay and Show (1992)
Industrial sludge	20.24	8.08	29.14	1.94	3.65	0.17	1.33	2.28	NA	9.33	1135	Show et al. (2006a)

Notes: [a] Ion chloride and lime dosing.
[b] Organic polymer dosing.
NA: data not available.

Table 2.5.   Chemical composition of sludge ash

Content	Range ($\times 10^{-1}$ %)
Silicon	80.1–141.0
Iron	80.0–81.3
Sulfur	62.4–110.9
Aluminum	42.3–55.9
Zinc	26.4–27.2
Calcium	25.0–31.0
Copper	18.4–19.3
Magnesium	7.7–8.1
Potassium	5.8–6.4
Chromium	2.8–3.2
Lead	2.4–2.8
Sodium	2.2–3.0
Nickel	1.2–1.6
Manganese	0.8–1.5

Table 2.6.   Physical properties of sludge ashes

Bulk density (kg/m³)	Absorption (%)	Specific gravity	Porosity (%)	Temperature (°C)	Reference
560–768	NA	NA	NA	1080	Elkins *et al.* (1985)
596–603	7.82–9.60	0.90–1.10	63.3–69.2	1080	Tay and Yip (1989)
480–780	2.49–19.16	NA	NA	1060–1100	Bhatty and Reid (1989b)
2,250	0.4	3.25	30.8	1135	Tay *et al.* (2000), Show *et al.* (2005, 2006a)

Note: NA: Data not available.

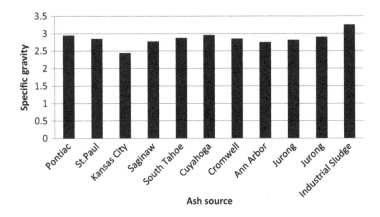

Figure 2.2.   Specific gravity of sludge ashes

Mineralogical compositions of sludge ash incinerated at 600°C are listed in Table 2.5. The ash mainly consists of Si, S, Fe, Al, Ca, Zn, Cu, and other trace metals. Most of the inorganic chemicals present in the sludge ash are not harmful to concrete, except for the high level of sulfur, which might lead to volume instability in the concrete matrix. The effects of melting processes on sludge ash properties had also been reported by Arakawa *et al.* (1984).

Summaries of physical properties of sludge ashes are given in Table 2.6 and Figure 2.2. As shown in Table 2.6, bulk density of the ash incinerated at temperatures ranging from 1060 to 1135°C falls within the range of 480–871 kg/m³ for municipal sludge and at approximately 2,250 kg/m³ for industrial sludge. Water absorption ranged from 2.49% to 19.16% for municipal sludge ashes and at approximately 0.4% for industrial sludge ash (Show *et al.*, 2005; Tay and Yip, 1989; Bhatty and Reid, 1989b). The municipal sludge ash was having a higher porosity of 63.3–69.2% comparing with 30.8% for that of industrial sludge ash. SG values of the ashes and their corresponding incineration temperatures are shown in Table 2.7 and Figure 2.2. The values of SG fall within a narrow range of 2.75–2.96, with the exception of Kansas City ash, which exhibits a low value of 2.44, and the industrial sludge ash with a high value of 3.25. The average SG value of the ashes is 2.85.

Table 2.7.   Effect of firing temperature on specific gravity

Ash source	Firing temperature (°C)	Reference
Pontiac	1400–1700	Gray and Penessis (1972)
St. Paul	—	Gray and Penessis (1972)
Kansas City	1500–1600	Gray and Penessis (1972)
Saginaw	1400–1700	Gray and Penessis (1972)
South Tahoe	1400–1600	Gray and Penessis (1972)
Cuyahoga	—	Gray and Penessis (1972)
Cromwell	—	Gray and Penessis (1972)
Ann Arbor	1600–1700	Gray and Penessis (1972)
Jurong	550	Tay (1987a)
Jurong	1050	Tay and Yip (1989)
Industrial sludge	1135	Tay *et al.* (2000), Show *et al.* (2005, 2006a)

Yip and Tay (1990) examined the effects of firing temperature on the SG of sludge as shown in Figure 2.3. It was found that beyond the pyroplasticity stage, the residues could achieve SG values higher than 2.80. Such SG values are comparatively higher than the average value of 2.65 for naturally occurring dense aggregate. This can be attributed to the residues containing the fusion of various metallic compounds when the pyroplasticity stage is reached. The effects of firing time on the SG values of sludge ash are shown in Figure 2.4. It was observed that firing beyond 8 h at the temperature of 1100°C has only marginal effects on the SG of the sludge ash. The SG of the incinerated sludge residue is 2.90, and with reference to Figure 2.3, the SG value is comparatively higher than that achievable under the furnace-firing conditions at 1080°C. This is due to the differences in the temperature history of the firing process.

Wang *et al.* (2008) investigated the sintering behavior of dried sewage sludge and the related sintering mechanisms. The experimental results indicated that the characteristics are primarily influenced by sintering temperature. When sintering temperature increased from 1020°C to 1050°C, the compressive strength and bulk density of the ceramic

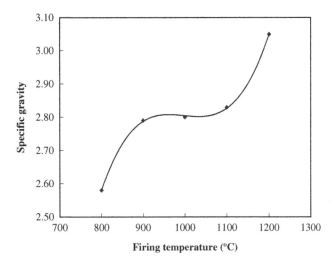

Figure 2.3.   Effect of firing temperature on specific gravity of incinerated sludge residue

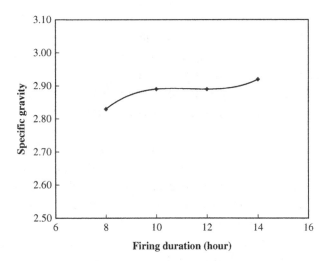

Figure 2.4.   Effect of firing time on specific gravity of incinerated sludge residue

specimens increased significantly while water absorption decreased, indicating an improvement of densification due to sintering. The compressive strength was unable to meet the requirement for traditional ceramic products due to the release of organic matters and the formation of large pores in the products. It was found that phosphorus in sewage sludge involved in the initial reactions with the formation of calcium magnesium phosphate and aluminum phosphate during sintering, which were helpful for enhancing the compressive strength. It was recommended that materials with high contents of aluminum could be used to enhance the compressive strength of products.

Sludge and sludge ash mainly contain the basic earth elements that were suitable for the production of various building materials. The properties and applications of building materials made from sludge and sludge ash are discussed in the following chapters.

3

Bricks

Bricks are generally made from clay and used in construction for aesthetic, economical and durability purposes. Brick-making materials are required to behave plastically when moistened while retaining its intended shape. Raw bricks ought to have sufficient "green strength" for handling after drying and should not crack or shrink excessively on drying or during firing.

3.1 Brick Production

Alleman and Berman (1984) produced "biobricks" using sludge with a solid content in the range of 15–25% mixed with clay and shale. Shale was mixed with clay at 1:2 ratio by volume and up to 50% by volume sludge. The materials were then uniformly blended in a commercial-grade mixer with appropriate amount of water to maintain the desired plasticity levels. The wet clay–shale–sludge mixture was then placed in the brick extrusion machine for molding. The freshly extruded bricks were then left to dry naturally and then dried in an oven at the temperature of 110°C for 36 h prior to firing. The "green" bricks were fired in a standard electric kiln at the temperature of 1100°C with temperature stair-stepped between 200°C. The stair-stepping technique was to prevent accelerated release of steam and gas within the brick body that could cause intolerable bloating, cracking, and deformation.

Tay (1987a) and Show and Tay (2004) demonstrated the use of both sludge and sludge ash from municipal wastewater, as brick-making

17

materials. A typical diagram of the production process is shown in Figure 3.1. For the sludge brick, a mixture of dried sludge and clay was ground and crushed into fine pieces by a crushing machine. The crushed mixture was extruded into brick samples. The bricks were dried in an oven at 100°C for 24 h and fired in a kiln at a temperature of 1000°C for about 24 h. For the sludge ash brick, the sludge was first fired in a furnace at 600°C and above, to remove the organic matter before grinding and crushing with the clay material, following the same production process.

Slim and Wakefield (1991) reported commercial production of bricks using municipal sludge and marine clay. Thirty percent by volume of sludge is used in the production of common bricks and between 5 and 8% is being used in face bricks. The sludge is blended into the clay and crushed to a size not exceeding 2.5 mm. The crushed materials are then mixed with water to achieve 20% moisture content and kneaded into a homogeneous mass for extrusion. The bricks are then air-dried and heated

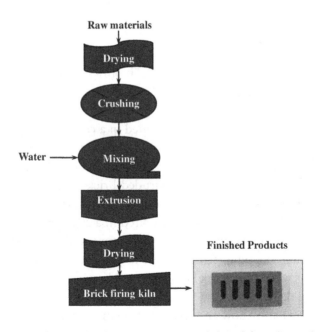

Figure 3.1.   Schematic brick production process (adapted from Tay *et al.*, 2004)

at a temperature of 150°C to remove the remaining moisture before firing at the temperature of 960°C. Trauner (1993) also demonstrated the use of sludge ash in brick production. The sludge ash was produced at a temperature of 870°C. As the sludge ash exhibited no plasticity, it was combined with moist clay to provide bonding during molding. The molded bricks were then dried at 105°C and fired at 1040–1120°C.

Okuno *et al.* (1997) reported commercial-scale production of 100% sludge ash bricks. However, due to the lack of plasticity of sludge ash material, stringent process requirements were stated. An average sludge ash particle size of less than 30 μm is required to prevent the development of hair cracks in the final products. The loss on ignition and moisture content in the sludge ash is required to be less than 1% each, to prevent cracks in the final product. It also requires a CaO concentration of less than 15% in the sludge ash to achieve a crack free appearance. The sludge ash also required high pressure molding at 1,000 kg/cm^2 in order to mold the ash without the use of additives, not even water. The bricks are then fired in a kiln at a rate of 600°C/h with a one hour holding time at the temperature of 900°C to complete the oxidation of organic substances, thus preventing the development of black core before the temperature is gradually increased to around 1000°C for sintering the sludge ash material.

Tay *et al.* (2004) reported the use of industrial sludge with marine clay in brick-making. The raw materials were dried at 105°C to achieve consistency in dryness, and ground to particle sizes finer than 0.25 mm to promote agglomeration. The bricks were made from 100% sludge as well as seven other mix proportions ranging from 0% to 70% marine clay with the remaining portion constituted of sludge. Initial trials indicated that the bricks with levels of marine clay above 70% developed cracks and distortions; therefore mixes with a clay content of over 70% were not investigated. The materials were added with 32% of water forming malleable pastes, which were extruded into bricks of dimension 32 × 40 × 75 mm. The raw bricks were dried and fired in a brick-firing kiln at 1050°C in a 24-h cycle, in which excessive moisture that can cause cracking were driven off gradually before subjecting to higher temperatures.

Huang *et al.* (2005) reported that sintering temperature requirement by the water treatment residual was higher than normally practiced in brick works due to the higher Al_2O_3 and lower SiO_2 content. The excavation waste soil, practically clay, was blended with water treatment residual to improve the brick quality. Under the commonly practiced brick-making condition, up to 15% of water treatment residual could be added to produce first grade brick specified by the local authority.

3.2 Properties of Bricks

Alleman and Berman (1984) showed that sludge can be used with clay and shale to produce "biobricks". In the bench-scale trial tests, "biobricks" containing 0–50% by volume of sludge have a density in the range of 1.63–2.09 g/cm³ (decreasing with increasing sludge content) and a water adsorption of 6.4–15.8% (increasing with sludge content). The compressive strength of the bricks ranged from 17.9 N/mm² with 50% sludge content to 35.1 N/mm² with no sludge content. It is reported that volumetric sludge addition of 25–30% is possible to produce bricks complying with ASTM criteria for severe weathering grade brick. Hence in the full-scale brick production, the bricks with 30% sludge addition exhibited a good compressive strength of 43.2 N/mm².

Tay (1987a) and Show and Tay (2004) compared the production of miniature bricks from both dried sludge and sludge ash. They found that the maximum percentage of dried sludge and sludge ash that could be mixed with clay for brick making are 40% and 50% by weight, respectively. The surface texture of the bricks was found to be uneven, mainly due to the organic component being burnt off during firing, and the degree of roughness increase with the amount of dried sludge in the brick. Addition of dried sludge reduced the specific gravities of the bricks from 2.38 for 0% sludge content to 1.98 for 40% sludge content. Addition of sludge ash, on the other hand, increases the specific gravities of the bricks from 2.38 to 2.58.

Water absorption values of bricks increased with the addition of both dried sludge and sludge ash from 0.03% to 3.63% and 1.70%, respectively. The shrinkage after drying is 4% for 0% sludge content as well as all percentages of dried sludge, and 3% for 50% sludge ash content. The shrinkage after firing ranged from 9.9% for 0% sludge content to 12.87% for 40% dried sludge, and 10.51% for 50% sludge ash content.

The compressive strength of bricks ranged from 87.2 N/mm^2 for 0% sludge content to 37.8 N/mm^2 for 40% sludge content, and 69.4 N/mm^2 with 50% sludge ash content (decreasing with increasing sludge content). It was observed that with addition of sludge ash in the production of brick, the strength was reduced while the specific gravity (SG) increased. However, bricks made with up to 50% sludge ash comply with the British Standard BS class 10 load-bearing bricks. It was reported that the addition of sludge ash beyond 50% would result in poor bonding of the materials during the extrusion. The addition of sludge ash was observed to increase water absorption and to reduce shrinkage.

Slim and Wakefield (1991) in the production of full-scale sludge bricks reported densities of 1.53 and 1.38 g/cm^3 for the 5–8% and 30% sludge addition, respectively. The bricks have drying shrinkages of 9.8% for 5–8% sludge addition and 11.1% for 30% sludge addition. The firing shrinkages were 4.3% and 3.6% for 5–8 and 30% sludge addition, respectively. The average water absorptions of the bricks were measured to be 13.1 and 14.2% for 5–8 and 30% sludge addition, respectively. The average compressive strength was measured to be 40.7 N/mm^2 for 5–8% sludge addition and 38.3 N/mm^2 for 30% sludge addition.

Trauner (1993) carried out a bench-scale production of bricks made with sludge ash and clay. The bricks have densities in the range of 1.88–1.50 g/cm^3 for firing at 1040°C and 1.94–2.12 g/cm^3 for firing at 1120°C (decreasing with increasing sludge ash content). The water absorption values were 13.25–22.54% (increasing with sludge ash content) for bricks fired at 1040°C and 4.5–1.2% (decreasing with increasing sludge ash content) for bricks fired at 1120°C. The compressive strength of the bricks ranged from 19 N/mm^2 with 30% sludge ash content to 46 N/mm^2 with no sludge ash added for the firing temperature of 1040°C and 73 N/mm^2 with 30% sludge ash content to 91 N/mm^2 with no sludge ash added for the firing temperature of 1120°C.

Okuno *et al.* (1997) produced 100% sludge ash bricks found that firing the bricks around the temperature of 1070–1080°C achieves the best quality bricks. The bricks can achieve a compressive strength of 200 N/mm², firing shrinkage ratio of 0.86 and water absorption of 0.1%.

Tay *et al.* (2004) produced bricks with industrial sludge and marine clay. They found that the drying shrinkage decreased with increasing sludge content, indicating a reduction in shrinkage rate with the addition of industrial sludge. The firing shrinkage on contrary increased with sludge content, as marine clay is prone bloating and expansion during firing. The addition of industrial sludge also increased specific gravity and reduced water absorption of the bricks. The brick made from 70% industrial sludge achieved a compressive strength of 31.4 N/mm², which is the highest among the various combinations. It was found that, bricks made from 100% sludge and 90% sludge with 10% clay, were prone to develop cracks during firing. Bricks of all mix proportions, other than 50% clay content, conform to the specified water absorption limit of 7%.

A comparison of results from the studies of producing bricks from sludge and sludge ash is presented in Table 3.1. It should be noted that the results from the study of Alleman and Berman (1984) were originally expressed in relation to percentage of sludge by volume. The expression of the results in relation to percentage of sludge by weight shown in Table 3.1(a) is based on the assumption that density of sludge and clay are 1,000 and 2,650 kg/m³, respectively for Alleman and Berman (1984), and the dry sludge density of 1,690 kg/m³ is used for Slim and Wakefield (1991).

The compressive strength of bricks incorporating sludge or sludge ash is shown in Table 3.2, while the strength of bricks made from industrial sludge is illustrated in Figure 3.2. It should be noted here that the considerable discrepancies between the studies could be attributable to the differences in testing methods, raw materials and sizes of the bricks used in the studies. The miniature bricks produced by Tay (1987a) and Tay *et al.* (2004) were tested according to British Standard BS 3921: 1985, whereas the bench-scale bricks produced by Alleman and Berman (1984) were tested to ASTM Standard C67.

Table 3.1. Properties of bricks made of sludge and sludge ash

Content*	Specific gravity		Water absorption (%)		Drying shrinkage (%)		Firing shrinkage (%)	
	Tay (1987a)	Alleman and Berman (1984)	Tay (1987a)	Alleman and Berman (1984)	Tay (1987a)	Slim and Wakefield (1991)	Tay (1987a)	Slim and Wakefield (1991)
(a) Bricks made of sludge								
0.0	2.38	2.09	0.03	6.4	4.0	NA	9.91	NA
10.0	2.32	NA	0.74	NA	4.2	NA	10.15	NA
20.0	2.24	1.76	1.37	13.6	3.7	11.1	10.84	3.6
30.0	2.17	NA	2.58	NA	4.2	NA	12.26	NA
40.0	1.98	NA	3.63	NA	4.0	NA	12.87	NA
	Tay (1987a)	Trauner (1993)	Tay (1987a)	Trauner (1993)	Tay (1987a)	Trauner (1993)	Tay (1987)	Trauner (1993)
(b) Bricks made of sludge ash								
0.0	2.38	1.88	0.03	13.25	4.0	3.4	9.91	1.0
10.0	2.42	1.69	0.07	18.80	2.5	2.1	9.95	1.1
20.0	2.46	1.56	0.11	19.37	2.5	1.2	9.10	2.1
30.0	2.50	1.50	1.39	22.54	3.4	0.5	9.36	3.5
40.0	2.55	NA	1.52	NA	3.2	NA	9.79	NA
50.0	2.58	NA	1.70	NA	3.0	NA	10.51	NA

(Continued)

Table 3.1. (*Continued*)

Content*	Specific gravity Tay et al. (2004)	Water absorption (%) Tay et al. (2004)	Drying shrinkage (%) Tay et al. (2004)	Firing shrinkage (%) Tay et al. (2004)
(c) Bricks make of sludge ash				
30	2.90	3.3	20.0	24.8
40	2.89	1.3	18.8	24.8
50	2.95	23.1	13.5	−11.2
70	3.13	1.7	16.1	46.3
80	3.14	0.9	16.5	41.1
90	3.32	1.4	13.8	43.5
100	3.34	1.5	12.0	48.4

Notes: *Percent sludge by weight. NA: Data not available.

Table 3.2. Compressive strength of bricks made of municipal sludge and sludge ash

Content*	Compressive strength (N/mm²)			
	Dried sludge	Sludge ash	Dried sludge	Sludge ash
0	35	47	87	87
3	30	NA	NA	NA
6	25	NA	NA	NA
10	—	30	60	85
14	23	NA	NA	NA
20	20	21	50	80
27	17	NA	NA	NA
30	NA	18	42	73
40	NA	NA	40	73
50	NA	NA	NA	70

Notes: *Percent sludge by weight. NA: Data not available.

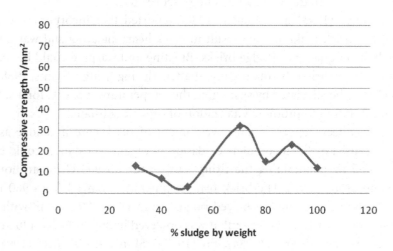

Figure 3.2. Compressive strength of bricks made of industrial sludge

3.3 Summary

Wastewater sludge and sludge ash can both be used in the production of masonry bricks. Sludge is usually not exceptionally high in plasticity and requires the addition of some clay material to provide a better cohesion.

Zakaria and Cabrera (1996) reported that addition of clay improves the performance of aggregates produced from fly ash materials. Sludge ash materials usually have no plasticity, and the addition of coarse, non-plastic materials to clay is known to reduce drying shrinkage of the clay products (Norton, 1970). The usual sludge incorporation ranged up to 40% by weight and the incorporation of sludge ash ranged up to 50% by weight.

Okuno *et al.* (1997) was able to produce bricks from 100% sludge ash without the use of any additives, however, stringent requirements were applied in the production procedures such as an average particle size of finer than 30 μm, moisture content of below 1%, and ignition loss of below 1% to prevent cracking. Alleman and Berman (1984) also reported that without careful control of firing temperature, accelerated release of steam or gas, or both, within the brick body would cause intolerable bloating, cracking, and deformation. Slim and Wakefield (1991) reported problems including distortion and cracking, and difficulty in the control of firing temperature with the addition of sludge above 45%.

Trauner (1993) and Adams (1988) reported that incorporation of sludge in brick-making often result in black heart, bloating and warping which affect quality of sludge bricks. Bloating and warping could be due to the gases released by the organic matters during firing (Adams, 1988), which may be alleviated by sustaining the temperature at 550°C for a sufficient duration to promote vitalization of organic substances.

It has been reported that development of black core in sludge ash bricks caused by the incomplete oxidation of organic substances could be prevented by maintaining the firing temperature at 900°C for an hour (Okuno *et al.*, 1997). The brick firing temperature ranged from 960 to 1120°C and the strength achieved ranged from 18 to 200 N/mm^2 with a density range of 1.37–2.00 g/cm^3. It was observed in the studies conducted by Alleman and Berman (1984), Tay (1987a), Slim and Wakefield (1991), and Trauner (1993) that the increase in sludge addition in bricks would cause reduction in strength and increase in water absorption of the brick product, attributable to higher porosity of the structure caused by the bloating effect. A comparison of some results from the brick production studies is presented in Table 3.3.

Table 3.3.   Summary of brick properties

Constituents	Alleman and Berman (1984)	Slim and Wakefield (1991)	Trauner (1993)	Okuno et al. (1997)
Constituents	Sewage sludge, clay, shale	Sewage sludge, clay	Sewage sludge ash, clay	Sewage sludge ash
Sludge input	15–50%v	5–30%v	10–30%w	100%
Temperature (°C)	1100	960	1040–1120	1020–1080
Strength (N/mm²)	43.2–45.8	38.3–40.7	18–80	80–200
Density (g/cm³)	1.63–1.94	1.37–1.53	1.5–2.0	NA
Water adsorption (%)	5–6	13.1–14.2	2.0–23.0	0.1–10.0
Shrinkage (%)	NA	13.7–14.3	9.0–11.5	~3–4

Note: NA: Data not available.

Addition of sludge to clay for brick-making generally reduces shrinkage but has no improvement on strength property and also increases the water adsorption. Application of sludge in brick-making can be a viable sludge disposal option which could be viewed as an opportunity for raw materials conservation and possible cost savings.

4

Concrete Aggregates

Aggregate is used in concrete as economic filler providing concrete with better dimensional stability and wear resistance. Aggregates made up 75% of concrete volume hence provide a great potential for large quantities of sludge application. However, sludge aggregates should not react adversely with other constituents of the mix, increase the cement requirement or have any adverse effects on the concrete properties, and should preferably provide beneficial value to the concrete.

4.1 Aggregate Production

Screened domestic wastewater sludge combined with clay, alum, and poly-acrylic acid was used to produce a lightweight building aggregate (Elkins *et al.*, 1985). The sludge was collected at the settling tank where it was thickened to 45% solids. The sludge material was further mixed with clay and pelletized. The pellets were fired at temperature between 1070 and 1095°C. The resulting product was a strong ceramic-like material.

Tay and Yip (1988) and Yip and Tay (1990) produced lightweight aggregate by firing digested and dewatered sludge samples in a brick-making kiln at the temperature of 1050°C for 6 h. The resulting sludge ash was then crushed and graded the desired particle size range, from 5 to 20 mm for use as coarse aggregates and 150 μm to 5 mm for use as fine aggregates. Tay *et al.* (1991) conducted a subsequent study on the production of lightweight aggregate by mixing digested and dewatered sludge collected from the sewage treatment plant with clay and subsequently fired the mixture in a brick-making kiln at 1050–1080°C. The ash produced was crushed and graded to the required aggregate sizes for use in concrete. A particle density contour plot with respect to raw sludge–clay mix and firing temperature (Figure 4.1) was established (Show *et al.*, 2006b). The contour plot provides a useful guide for production of lightweight aggregates in terms of firing temperature and raw material mix proportion.

Bhatty and Reid (1989a) demonstrated the use of sludge ash in the production of lightweight coarse aggregate in both pelletized and crushed forms. Due to the lack of plasticity in sludge ash, the particle

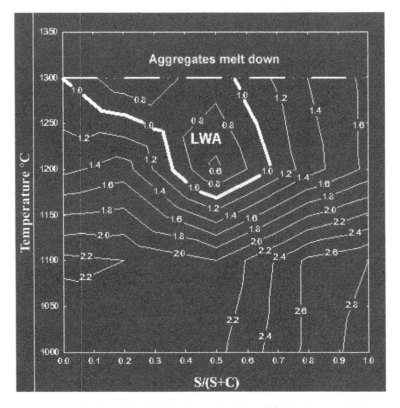

Figure 4.1.   Contour plot of sludge–clay mass as S/(S+C) and firing temperatures on particle density (g/cm^3). LWA: region of lightweight aggregates (adapted from Show *et al.*, 2006b)

size is required to be exceptionally fine in order to bind without the use of additives. Sludge ash that is available in fine powdery form with a mean particle size of 35 μm was used. It was also reported that the ash particle size distribution matches the range for natural pelletizing as suggested by Morinaga *et al.* (1963). The ash was screened through an 850-μm sieve to remove lumps and moistened with 20% water. The moist ash was then fed into a laboratory balling unit consisting of 50.8 cm tire rotating at 45 rpm at an angle of 20° to produce round sludge pellets. Moist ash was also prepared into layers of 6 mm thickness to produce sludge slabs. The processed sludge ash was then dried overnight at a temperature of 100°C to complete dryness before firing in a muffle furnace.

It was found that porosity of sludge ash increases with temperature up to 1090°C when melting begins and results in a denser structure. The sludge pellets were fired at temperatures ranging from 1060°C to 1085°C for a duration of 10–30 min while the sludge slabs were fired at temperatures ranging from 1070°C to 1075°C for 15–20 min. The sludge slabs were later crushed in a roller crusher to obtain a desired size range while the pellets can be applied directly in concrete.

Bhatty *et al.* (1992) demonstrated the use of sludge ash in the production of both crushed and pelletized coarse normal weight aggregate for use in concrete. The particle size of the sludge ash material required a size range finer than 100 μm with mean below 50 μm to be suitable for molding at a moisture content of about 25%. The production process used is similar to the process mentioned in Bhatty and Reid (1989a). Aziz and Koe (1990) also conducted investigations on the production of mix-grade lightweight aggregates from burnt sludge and marine clay. A range of aggregates were produced using 20% and 30% burnt sludge and fired with clay at the temperature of 1000–1100°C. Celik and Bayasi (1995) conducted a comparison of the performance of sludge–clay aggregates produced with the procedures described by Elkins *et al.* (1985) using a firing temperature of 1077–1088°C.

Tay and Show (1999), Tay *et al.* (2000, 2001a,b, 2003) and Show *et al.* (2006a) conducted studies on the use of industrial sludge that contains little organic materials in the production of regular weight artificial aggregates. Flow diagram of the production process of sludge–clay concrete aggregates is shown in Figure 4.2. The raw sludge and clay were dried at a temperature of 105°C and pulverized separately to a suitable fineness for mixing, using a mechanical pulverizer. As raw industrial sludge and marine clay has better cohesion, particle sizes of below 150 μm is sufficiently fine for subsequent mixing and molding when added with water ratio of 35–42%. Energy dispersive X-ray (EDX) spectroscopy was used as a chemical microanalysis technique in examining the extent of the material mixing. Used in conjunction with scanning electron microscopy (SEM), the EDX Spectroscopy technique detects X-rays emitted from the sample during bombardment by an electron beam to characterize the elemental composition of the analyzed volume. Figure 4.3 demonstrates typical EDX spectra of the interior of 20:80 sludge–clay specimen sintered at 1200°C.

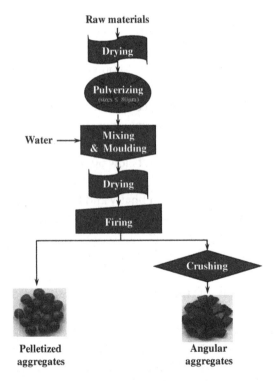

Figure 4.2.   Schematics of sludge–clay aggregates production (adapted from Show *et al.*, 2006a)

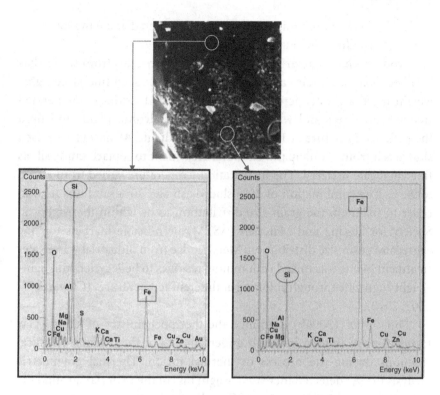

Figure 4.3.   EDX spectra of the interior of 20:80 sludge–clay specimen sintered at 1200°C (adapted from Show *et al.*, 2006b)

The energy spectrum indicates chemical compositions from two specific spots within the sintered aggregate at a distance of 100 micron. As shown in Figure 4.3, the compositions differed considerably, particularly for Fe and Si. Given Fe and Si as the main constituents of sludge, and those of clay were Si and Al, the non-homogeneity of the compositions indicated incomplete mixing of the two raw materials.

After mixing, the mixtures were either pelletized into spherical shapes for direct applications after sintering, or pressed into flat layers which were subsequently crushed after sintering to produce angular aggregates. The first firing ramp was held at 550°C for an hour to ensure complete removal of organic matters. It was then raised to 900°C and held at this tempera-ture for another hour. The 1-h residence time at 900°C is to ensure complete oxidation of the mixtures in order to prevent the development of black core. The fired mixtures were finally sintered at a temperature of 1135°C to produce the final products.

Production of fine artificial lightweight aggregates from sludge has also been explored. Kato and Takesue (1984) produced fine-sized light-weight aggregate for concrete by mixing pulverized residue of incinerated sludge with water and waste liquor from an alcohol plant, and fired the pelletized mixture at 1050°C. Khanbilvardi and Afshari (1995) used sludge ash from municipal wastewater treatment to replace sandy silt as fine aggregate material. The incineration temperature varied from 700°C to 90°C. The combination of 30% sludge ash and 70% sand was used in order to maintain the grain size distribution to be within the American Society for Testing and Materials (ASTM) permissible margins for fine aggregates grain size distribution. Metal sludge from industrial wastewater treatment plants was mixed with mining residues to be recycled into light-weight aggregate through sintering at different temperatures (Chang *et al.*, 2007, 2010).

Show *et al.* (2006b) examined the thermochemistry of sludge–clay mixes in providing an in-depth understanding of the thermal-chemical reactions during the process of thermal fusion. Thermal gravimetric analysis (TGA) detected increased weight loss of the materials around the temperatures of 520°C and 900°C, indicating volatilization of organic matter and inorganic salts, at the two respective temperature levels. Differential scanning calorimetry (DSC) results revealed endothermic reactions possibly due to dehydroxylation of clay between temperatures of 480°C and 600°C, and calcite decomposition between the tempera-tures of 680°C and 900°C. Intensified endothermic peaks were observed in the mixes of the two materials.

TGA-Infrared revealed increased intensities of the absorption bands at 1,600 cm^{-1} at 630°C and 2,900 cm^{-1} at 710°C, as a result of the break down of organic matter and alkanes, respectively. Energy demand analysis deter-mined the energy required for raising the temperature of each mix to reach each specific temperature. The intrinsic energy required to raise the firing temperature of the aggregates from 400°C to −900°C were deter-mined to be within the range of 20,347 and −38,254 kJ/kg, with 100% sludge exhibiting the lowest energy requirement and 20% sludge exhibiting the highest energy requirement.

Rao *et al.* (2009) investigated the reuse of fluoride contaminated bone char sludge as partial replacement for fine aggregate in the manufacture of dense concrete specimens. The impact of admixing 1.5–9% sludge contents on the compression strength and fluoride leaching potential of the sludge admixed concrete (SAC) specimens were examined with respect to strength criteria for manufacture of dense, load-bearing concrete blocks.

Washing aggregate sludge from a gravel pit, sewage sludge from a wastewater treatment plant and a clay-rich sediment were mixed, milled and formed into pellets, pre-heated and sintered in a rotary kiln at 1150, 1175, 1200, and 1225°C for 10 and 15 min at each temperature to produce aggregates (Gonzales-Corrochano *et al.*, 2009). All the mixtures presented a bloating potential resulting in lightweight aggregates in accordance with Standard UNE-EN-13055-1. Three groups of lightweight aggregates on the basis of their properties in comparison to commercially available lightweight aggregates were established with similar applications such as horticulture, prefabricated lightweight structures and building structures.

4.2 Properties of Aggregates and Concrete

Elkins *et al.* (1985) produced an expanded lightweight aggregate with a bulk loose density between 560 and 768 kg/m^3. The lightweight concrete produced with the aggregates exhibited a unit weight of 1,681–1,825 kg/m^3 and a 28 day concrete strength in excess of 35 N/mm^2. Incinerated sludge ash was found to have a specific gravity (SG) of 2.90 which is comparatively higher than the range of 2.60–2.70 for ordinary dense aggregates (Tay and Yip, 1988, 1989). Despite high SG, it has a high mean porosity of 66%, provides a characteristic of low unit weight that is typical of lightweight aggregates. The effect of burning duration on the specific gravity of the sludge ash was found to be marginal after 8 h, while the specific gravity increased notably when the firing temperature is raised, especially beyond 1100°C (Yip and Tay, 1990).

The mean air-dry bulk densities of the fine and coarse sludge ash aggregates were determined to be 900 and 600 kg/m^3, respectively. The water absorption of the sludge ash is lower at 8.5%, compared to the usual

19–20% for foamed slag and expanded clay, respectively. The low water absorption values could be due to the large amount of metal compounds present resulting in a poor affinity for moisture. The kiln-incinerated sludge residue fired at the temperature of 1080°C obtained a 10% fine value of 34 kN which compares well with the values of 31 and 32 kN for other commercial lightweight aggregates. The 28-day compressive strengths of the lightweight aggregate concrete ranged from 7 to 20 N/mm^2, depending on the mix design (Tay and Yip, 1989).

The low thermal conductivity and high fire resistance of the sludge ash aggregates may render them suitable for use in thermal insulation and fire protection of concrete. The lightweight concrete beam retained 89% of its ultimate strength, while the corresponding value for the ordinary dense concrete is 80%, showing that the sludge ash lightweight has better fire resistance than ordinary dense concrete (Tay and Yip, 1988). The concrete made of sludge ash aggregates was found to have a remarkable fire resistance that is superior to that of the ordinary concrete.

Physical properties of incinerated clay-blended sludge coarse aggregate produced by Tay *et al.* (1991, 2002a, 2004) are given in Figure 4.4. The particle density, 10% fines, and bulk density of the aggregate show an increasing trend as the clay composition increases from 10% to 40%. On the other hand, the specific gravity, water absorption, and porosity decrease as the percentage of clay increases from 10% to 40%. The 28-day compressive strengths of the lightweight aggregate concrete are given in Figure 4.5. As shown in Figure 4.5, the 28-day compressive strength increases from 10% to 40% clay for all mixes. The results also indicated that the optimum fine-coarse aggregates ratio at which maximum compressive strengths occur is 1:1.50. At this optimum ratio, the compressive strength increases from 23.9 N/mm^2 for 10% clay to 31.0 N/mm^2 for 40% clay. British Standard BS 8110-1:1985 recommends a minimum characteristic crushing strength of 15.0 Nmm^2 at 28 days for reinforced lightweight aggregate concrete. The recommended 28-day strength of 15.0 N/mm^2 is attained by all incinerated clay-blended sludge lightweight aggregates. Results from the study by the authors indicated that incinerated clay-blended sludge is a potential material for the production of lightweight aggregate concrete for structural application.

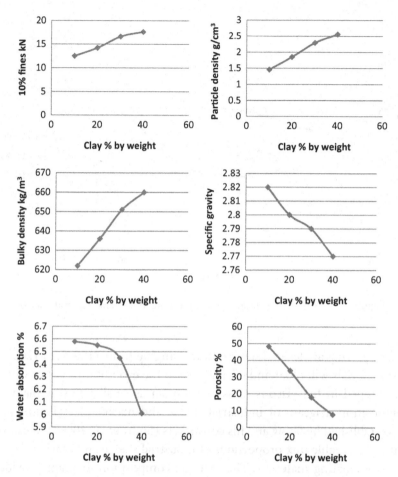

Figure 4.4.   Physical properties of clay-blended sludge aggregates

Bhatty and Reid (1989a) mentioned that two conditions essential for expansion of such materials to occur were suggested by Riley (1951), Ehlers (1958), and Utley *et al.* (1985) to be: the development of a glassy phase over a wide temperature range, and evolution of gases from the dissociation of mineral components such as carbonates, oxides, hydrates, and sulfates present in the starting material to form cellular structure in the mass. Lightweight aggregates are usually produced from clay, shale and

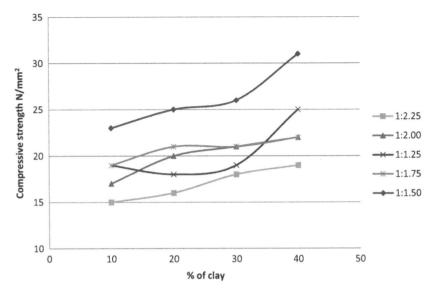

Figure 4.5.   28-day concrete strength with varying fine/coarse aggregate ratios

slates by firing at elevated temperatures ranging from 1100°C to 1200°C. The materials will bloat and expand due to the evolution of gasses which are trapped in the viscous mass and remain as small pores. A material composition diagram of industrial sludge and marine clay (Figure 4.6) was established by Show and his co-workers (Show *et al.*, 2006b) in examining the suitable mix proportions of industrial sludge and marine clay as the raw bloating materials. The material composition diagram provides useful information of material bloating with respect of the mix compositions upon firing of the raw mixtures.

Bhatty and Reid (1989a) found that the pellets fired between 1060°C and 1100°C for periods of 10–30 min showed lower specific gravity at higher firing temperature and longer firing duration. This is due to an increase in the porosity of pellets as a result of gases released from the sludge ash at high temperatures. The specific gravity of the pellets continues to fall with the increase in temperature as porosity rises until the temperature of 1090°C when melting begins and the pores start to collapse resulting in a slight increase in specific gravity. Overexpansion may also occur when pellets are heated beyond 1090°C for longer periods.

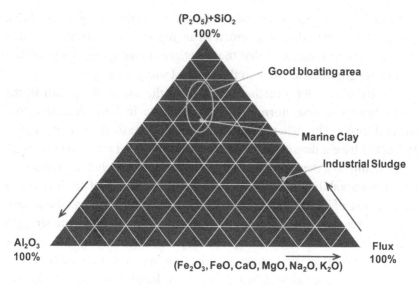

Figure 4.6. Material composition diagram of industrial sludge and marine clay (adapted from Show *et al.*, 2006b)

Overexpanded aggregates were weaker due to larger bubbles and thinner skin. It was found that pellets with a uniform expansion of specific gravity form 0.8 to 1.4 are the most suitable for use as lightweight aggregates and can be produced within the temperature range of 1060–1085°C. Pellets of specific gravity between 1.1 and 1.4 can achieve concrete compressive strength of up to 37 N/mm² and pellets of specific gravity between 0.8 and 1.1 can achieve concrete compressive strength of up to 24 N/mm².

In the sludge slabs experiment, the specific gravity also falls with increasing firing temperatures until after the temperature of 1090°C when melting takes place and a denser structure results. The temperature selected for firing the sludge slabs are 1070°C and 1075°C with the duration of 15 and 20 min, respectively. The two groups of aggregates can achieve concrete compressive strengths of 28 and 22 N/mm², respectively. The aggregate strengths are good compared to commercial expanded clay which has a concrete compressive strength of 29 N/mm². The artificial sludge ash pellet aggregates have substantially lower absorption values of 2.49–2.88% due to their impervious skin, whereas the crushed sludge slabs have higher absorption values of 9.16–19.36% due to their open pore surfaces. The aggregates provide concrete strengths of up to 37 MPa, which is superior to that of commercial aggregates derived from expanded clays. Results also indicated that the pelletized sludge aggregates provide a better strength in concrete than the crushed sludge aggregates.

Bhatty *et al.* (1992) further investigated the use of sludge ash in the production of coarse normal weight aggregate in both pelletized and crushed forms. The pellets produced from sludge ash at the temperature of 1050°C have a density of 1.77 g/cm^3, and provided a concrete strength of 35 N/mm^2 at a density of 2,124 kg/m^3. Bhatty *et al.* (1992) also observed enhance workability and compaction in the properties of fresh concrete with the use of sludge pellets. The sintering of sludge ash has the advantage of developing inert aggregates of low density and adequate strength to be substituted for concrete aggregates, providing beneficial effect to the concrete without causing excessive reduction in concrete strength.

The mix-grade lightweight aggregates produced from burnt sludge and marine clay reported by Aziz and Koi (1990) displayed specific gravities ranging from 2.25 to 2.36, bulk density ranging from 910 to 1,025 kg/m^3, and water absorptions from 13% to 20%. The aggregates provide concrete compressive strengths ranging from 19.6 to 32.2 N/mm^2, with aggregates containing 5% cement addition achieving the highest strength. The compressive strength of the concrete made of pellet aggregates was also found to be higher than those of concrete made of crushed aggregates.

Tay and Show (1999) and Tay *et al.* (2000, 2001a,b, 2002b, 2003) reported the use of industrial sludge as regular weight artificial aggregates. The properties of the sludge–clay aggregates at various mix proportions are given in Figure 4.7. The aggregate strength was determined using aggregate impact value (AIV) test in British Standard for Aggregate testing (BS 812 Part 112: 1990). AIV is a measurement of the susceptibility of the aggregates to crushing; therefore a lower value denotes a better aggregate quality. The aggregates made from industrial sludge displayed moderate strength with the highest being comparable with that of normal granite, and decreases with increasing clay content. However, unlike granite aggregate that weakened upon wetting, the strengths of the sludge–clay aggregates improved upon wetting. The improvement in strength could be due to the pore water pressure effect brought about by the highly porous nature of the aggregates. The improvement was evident that even the

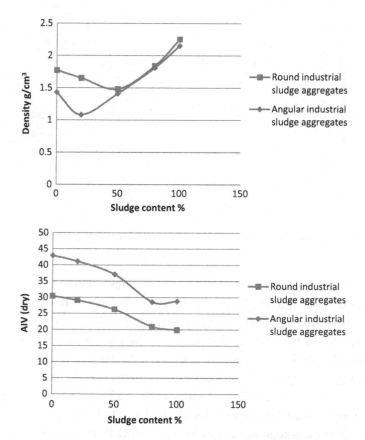

Figure 4.7. Properties of industrial sludge aggregates

crushed angular aggregates made from sludge and clay that were weaker compared to their round counterparts were all better in impact resistance than the granite aggregates in the wet condition.

The compressive strengths of the concrete made from the aggregates were tested and the results are given in Figure 4.8. Concrete samples cast from the sludge–clay aggregates yield compressive strengths in the range of 31.0–39.0 N/mm^2. The round sludge aggregates with 0% clay content and the angular sludge aggregates with up to 20% clay content are suitable for paving and structural applications while the rest of the aggregates could be used for other general applications where strength is not a critical requirement. The aggregates made from industrial sludge manifested

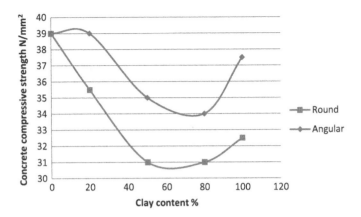

Figure 4.8.   28-day concrete strength with sludge–clay aggregates at varying clay content

the attributes required of construction aggregates, and the incorporation of marine clay reduced the particle density of the aggregates and concurrently the impact resistance of the aggregates.

Celik and Bayasi (1995) demonstrated the production of incinerated clay-blended sludge aggregate by firing at 1077–1088°C for use as coarse lightweight aggregate. The aggregates exhibited a specific gravity of 1.59 with water absorption of 7%, which provides a 28-day concrete compressive strength of 26 N/mm^2 at a density of 1,842 kg/m^3. It was found that the use of these sludge aggregates increases workability, permeability, and the drying shrinkage of the concrete.

In the use of fine artificial lightweight aggregates from sludge, Kato and Takesue (1984) were able to produce aggregates with a specific gravity of 1.33 which is lower than that of sand (2.61) as well as conventional lightweight materials (1.63). The aggregates achieved a 28-day concrete compressive strength of 41.0 N/mm^2 which is comparable to those achieved by sand and conventional materials at 46.9 and 44.8 N/mm^2, respectively. Khanbilvardi and Afshari (1995) in the study of using sludge ash as a portion of fine aggregates in concrete indicated that replacement of fine aggregate by sludge ash up to 30% by weight is possible in normal practice. The 28-day compressive strength of the concrete decreases as the content of sludge ash in the mix increases. The concrete strength was 21.9–24.3 N/mm^2 for 10–30% replacement values. The strengths are lower compared to the control of 28.5 N/mm² but still satisfied most design requirements.

Cheeseman and Virdi (2005) and Kosior-Kazberuk (2011) suggested incorporating sintered sludge ash into manufacturing of lightweight aggregate. The ash was mixed with a clay binder, formed into spherical pellets under 1020–1080°C. From the investigation, sintered sludge ash complied with the fundamental properties such as density, water absorption, and compressive strength superior to those commercially lightweight aggregate.

Show *et al.* (2006a) investigated the use of clay mixed with industrial sludge to produce high strength concrete aggregates in pelletized or crushed forms. The pelletized aggregates displayed a higher particle density ranged between 1.48 and 2.25 g/cm³ (Figure 4.7), compared to the crushed counterparts that had a particle density that fell between 1.08 and 2.15 g/cm³. Both types of aggregates displayed lower densities relative to that of granite with 2.56 g/cm³. The pelletized aggregate exhibited moderate aggregate strength of 18.1–30.4% AIV, while the crushed aggregate exhibited lower strength of 23.1–42.9% AIV, which could be caused by the process of crushing in which the aggregate strength may be impaired. The AIV values are comparable to that of granite, which ranged from 28.3 to 38.9%. A complete replacement of regular granite aggregates in concrete was accomplished by substituting the coarse aggregates with the same volume of sintered sludge–clay aggregate. The pelletized and crushed aggregate made from 100% sludge provided moderate 28-day concrete compressive strengths of 38.5 and 39.0 N/mm², respectively, which was comparable to that of the control granite specimens with strength of 38.0 N/mm². Aggregates of other sludge–clay combinations provided 28-day concrete compressive strengths ranging from 31.0 to 39.0 N/mm².

Kuo *et al.* (2007) suggested using the organo-modified reservoir sludge as a substitute of fine aggregate in cement mortar production. The experimental result showed that the compressive strength of cement mortar decreased as the percentage of sludge particles increased. The compressive strength of cement mortar at 28 days of 30% sludge was still higher than 30 MPa, while the strength with 30–50% sludge could be used for non-structural product such as brick and partitions. The replacement of sludge particles ranging from 5% to 50% in cement mortar improved the waterproofing performance.

In another study, dry sewage sludge (DSS) as the principal material was blended with coal ash (CA) to produce lightweight aggregate (Wang *et al.*, 2009). The effects of different raw material compositions and sintering temperatures on the aggregate properties were then evaluated. In addition, an environmental assessment of the lightweight aggregate generated was conducted by analyzing the fixed rate of heavy metals in the aggregate, as well as their leaching behavior. The results indicated that using DSS enhanced the pyrolysis-volatilization reaction due to its high organic matter content, and decreased the bulk density and sintering temperature. However, the sintered products of un-amended DSS were porous and loose due to the formation of large pores during sintering. Adding CA improved the sintering temperature while effectively decreasing the pore size and increasing the compressive strength of the product. Furthermore, the sintering temperature and the proportion of CA were found to be the primary factors affecting the properties of the sintered products, and the addition of 18–25% of CA coupled with sintering at 1100°C for 30 min produced the highest quality lightweight aggregates. In addition, heavy metals were fixed inside products generated under these conditions and the As, Pb, Cd, Cr, Ni, Cu, and Zn concentrations of the leachate were found to be within the limits regulatory requirements.

Latosinska and Zygado (2011) reported that sewage sludge can be used as an expanding agent in the production of lightweight expanded clay aggregate (LECA) mass. The addition of sewage sludge increases the total porosity and decreases the bulk density of a sinter. The addition of sewage sludge to a raw material used in the production of LECA enabled a decrease in the burning temperature for the maintained operational parameters of a lightweight aggregate. The optimum content of sewage sludge added to a raw material used in LECA production was 5–15% of dry mass. The addition of sewage sludge in an amount of 5% and 10% caused an increase in closed porosity.

4.3 Summary

Concrete aggregates had been made from both sludge and sludge ash and the density may vary depending on the nature of the sludge and the firing process used in the aggregate production. Sewage sludge more often has to be incinerated to destroy any pathogen present in the sludge as a health safety precaution prior to handling. Sewage sludge is therefore commonly incinerated and crushed for direct application as concrete aggregates. If molding and/or pelletizing is required, the incineration of sewage sludge would reduce the cohesion of the sludge material; hence the particle size would have to be sufficiently fine to be molded without the use of additives. Industrial sludge on the other hand, contains low level of organic contaminants and hence can be used directly in the raw state, providing a better adhesion for molding. Clay material is often mixed with sludge to improve the cohesiveness of the materials (Latosinska and Zygado, 2011).

The lightweight property can be achieved when the sludge is heated to a viscous state where the evolution of gases from the dissociation of mineral components can be trapped to form intercellular structure in the mass. The common temperature required for sewage sludge to achieve lightweight property ranges from 1050 to 1100°C. Higher temperature is required for materials such as clay and industrial sludge that contain lower amount of organic matter. The artificial sludge aggregates can provide concrete with compressive strengths in the range of 7–41 N/mm^2 depending on the properties of the aggregate and the amount of cement used in the concrete batching.

5

Cement Replacement Materials

Cement is the important binding agent that enables the formation of concrete. Cement material may consist of three following components; the hydraulic component, pozzolanic component and the inert component. The hydraulic component reacts with water forming cement gel that hardens and binds things together. The pozzolanic material is mainly composed of silicium and alumina oxides and reacts with lime to form a precipitate that provides the binding property. Inert materials that do not react with the cement materials can be used as fillers to improve the strength of cement while reducing the overall cost.

5.1 Cement Filler

5.1.1 Production of cement filler

Use of sludge ash as a partial replacement for cement in concrete has been reported (Tay, 1987b, 1987c). Sludge from sewage treatment works was treated at a temperature of 550°C to remove organic matter. The sludge ash produced was pulverized and particles finer than 1.15 mm were used as cement filler. Concrete cubes cast in molds of 100 mm × 100 mm in dimensions were used to study the effect of sludge ash on concrete strength.

Bhatty and Reid (1989a), investigates the potential for using sludge ash produced from the incineration of municipal sewage sludge as a fine aggregate in mortar. The sludge ash used was directly from the incineration plant with a mean particle size of around 50 μm. The addition of fly ash and lime were also examined to enhance any potential pozzolanic activity of the sludge ash. The strength characteristics of mortar prepared from sludge ash with the addition of fly ash and varying amounts of cement and lime were determined.

Monzo *et al.* (1996, 1999) carried out a study on the replacement of Portland cement with sewage sludge ash. The sludge ash used exists in the form of fine powder with a particle size of around 80–40 μm. The ashes were then sieved into different size fractions used in different fineness and different proportions in the preparation of cement mortar. Mortar samples containing sewage sludge ash were prepared replacing 15% or 30% of cement in control mortar and cured at 40°C.

Use of sludge ash as a sand or cement replacement in cement-based products has been investigated (Cyr *et al.*, 2007). A higher requirement of water was reported due to irregular surface of the ash. The sludge ash also reduced the workability of cement and prolonged the setting time. Although sludge ash-based mortar showed a lower compressive strength of 25–50%, their pozzolanic activity indicated a positive material strength.

Chen and Lin (2009) mixed incinerated sewage sludge ash (ISSA) with cement in a fixed ratio of 4:1 for use as a stabilizer to improve the strength of soft, cohesive, subgrade soil. Five different ratios (in wt.%: 0, 2, 4, 8, and 16%) of ISSA/cement admixture were mixed with cohesive soil to prepare soil samples. Tests of pH value, Atterberg limits, compaction, California bearing ratio (CBR), unconfined compressive strength, and triaxial compression were performed on the samples. The study showed that the unconfined compressive strength of specimens with the ISSA/cement addition was improved to approximately 3–7 times better than that of the untreated soil. The swelling behavior was also effectively reduced as much as 10–60%. For some samples, the ISSA/cement additive improved the CBR values by up to 30 times than that of untreated soil. The study suggested that ISSA/cement has many potential applications in the field of geotechnical engineering.

5.1.2 *Properties of filler and concrete*

The analytical results on the chemical compositions of sludge by Tay (1987b, 1987c) revealed that the inorganic compounds were chemically inert in nature. Effects on the segregation, shrinkage and water absorption of concrete were not significant with up to 40% of pulverized sludge ash. Workability was improved with an increase in amount of cement replaced by sludge ash above 15%. Setting times of the concrete samples with pulverized sludge ash were longer but still within the requirements of British Standard BS 12: 1978. The compressive strength of concrete (Figure 5.1) incorporating sludge-ash-blended cement decreased as the percentages of ash in the cement increased. However, the 28-day compressive strength of concrete cubes with 10% sludge ash in the cement was about the same as the control strength. For 40% sludge-ash-blended cement, the strength fell by about 50%.

Bhatty and Reid (1989b) suggested that the mortar made from sludge was unable to attain the strength of normal mortar due to the finely dispersed and porous nature of the ash. Despite its high silica, alumina and lime contents, the material lacks pozzolanic activity. Results indicate that the addition of sludge ash adversely affects the strength behavior of mortars. Although partial substitution of sludge ash with fly ash shows an

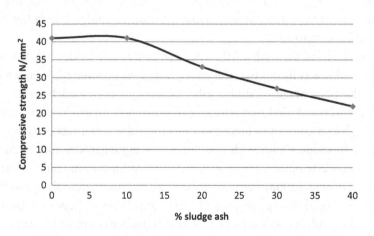

Figure 5.1.   28-day concrete strength with sludge ash filler

improvement in strength and workability, sludge ash mortar was not able to achieve strength comparable to that of sand mortar. The study carried out by Monzo *et al.* (1996) shows an enhancement of 28-day compressive strength of up to 15% can be achieved at a compressive strength of 47.7 N/mm², when ashes are used. It was however established that sludge ash has a negative influence on the workability of mortar mixes, due to the irregularity of the particle shape.

Monzo *et al.* (1999) carried out further investigation on the effect of sludge ash fineness and proportion, and the effect of cement types. In the examination of the gain of strength produced by the partial replacement of cement by sludge ash compared to a hypothetical behavior of mortar made with inert product of the same proportion, it was found that sludge ash behaves as an active material. It is evident in the sets of experiments conducted that the replacement of cement by sludge ash produces a positive relative compressive strength gain of 10–70%, which is more pronounced when the amount of cement replaced is higher. The increase of compressive strength compared to the control mortar indicated possible pozzolanic properties of the sludge ash. The high sulfur content in sludge ash mortars ranged from 3.78% to 6.28% did not have any major influence on the compressive strength of the mortar containing the sludge ash. In the comparison of grain size fraction, finer fraction of sludge ash provided a higher relative compressive strength gain even though it has a higher SO_3 content and a lower SiO_2 content, indicating that fineness is a more dominant factor for pozzolanic reaction in sludge ash rather than chemical composition.

The flexural strength of the mortars containing sludge ash was also tested, and it was found that the increase in flexural strength is enormous when the percentage of sludge ash replacement increased from 15% to 30%. It was concluded that sewage sludge ash is compatible with cements with high C_3A content as binder in mortars and it is not recommended to be used in conjunction with fine carbonate aggregates as longer curing times, delay ettringite formation or thaumasite formation could occur.

Garcia *et al.* (2008) investigated the use of paper waste sludge as pozzolanic material to replace cement. The conversion of paper sludge into pozzolanic additive was achieved by heat transformation of kaolinite into metakaolinite at 700°C for 2 h. It was reported that replacement of cement with 10% pozzolanic additive derived from paper sludge improved the compressive strength of concrete by up to 10% satisfying the chemical and mechanical criteria in accordance to European Standard UNE EN 197-1.

5.2 Masonry Binder

5.2.1 *Production of masonry binder*

The potential for utilizing digested and dewatered sludge to produce cementitious materials have been examined by Tay and Show (1991, 1992a, 1993). In preparing the cement specimens, sludge was oven-dried and mixed with limestone powder at various proportions by weight through the process depicted in Figure 5.2. The mixtures were ground and inciner-ated in a furnace at different temperatures and for different durations of controlled burning. The ash collected was ground to less than 80 μm

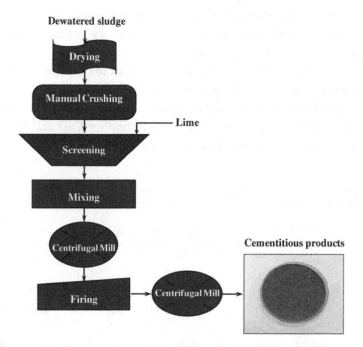

Figure 5.2. Schematic of sludge cement production process (adapted from Tay and Show, 1993)

before tested for various properties. In another study, sewage sludge was solidified/stabilized by a binder constituted of Portland cement and stone material powder (Wang and Liao, 2011). Based on the experiment data, a mathematical model relating the strength gain of solidified sludge matrices to the ratio of binder to waste is proposed by statistical analysis. The proposed method is then verified with available test data.

5.2.2 Properties of binder and masonry

An optimum condition of the mix composition, burning temperature burning duration and curing condition upon highest possible strength development of the cement was determined from the study. It was found that under air curing, the cement specimens with 50% sludge and 50% limestone, fired at 1000°C for 4 h exhibited the highest compressive strength. The cement specimens were tested for chemical, physical and compressive strength properties.

Chemical composition of the sludge cement is listed in Table 5.1. As a comparison, the chemical composition of ordinary Portland cement is also given. The limiting values (Labahn, 1983) stated in Table 5.1 are to be regarded as valid for the manufacture of cement for general works. The four major oxides of the sludge cement are SiO_2, CaO, Al_2O_3, and Fe_2O_3. Most of the chemical compositions of the sludge cement listed in Table 5.1 are within the limiting values, with the exception for CaO and SO_3 contents. The CaO contents are low whereas the SO_3 levels are excessive.

Table 5.1. Oxides in percent by weight (adapted from Tay and Show, 1991)

	SiO_2	CaO	Al_2O_3	Fe_2O_3	K_2O	MgO	Na_2O_3	SO_3	Loss-on-ignition
Portland cement	20.86	63.30	5.67	4.11	1.21	1.04	0.17	2.11	1.91
Sludge ash	20.33	1.75	14.64	20.56	1.81	2.07	0.51	7.80	10.45
Sludge cement	24.55	52.11	6.61	6.26	1.05	2.07	0.17	4.88	0.30
Limiting value	18–24	60–69	4–8	1–8	<2.0	<5.0	<2.0	<3.0	<4.0

Table 5.2.   Physical properties of cement

	Fineness (m²/kg)	Soundness (mm)	Bulk density (kg/m³)	Specific gravity	Consistency (%)	Pozzolanic activity index (%)	Initial set (min)	Final set (min)
Sludge cement	113	1.9	685	3.33	82	67.2	40	80
Ordinary Portland cement	116	0.9	866	3.16	27	100	180	270

Physical properties of the sludge cement are given in Table 5.2. The apparent bulk density was 685 kg/m³. The specific gravity value was 3.33. The sludge cement is rated sound since the soundness test result of 1.9 mm is well within the limiting value of 10 mm specified by British Standard BS 12: 1978. Based on the pozzolanic activity index result of 67.2%, the cement specimens exhibit low pozzolanic activity. A high water demand property of the sludge cement is indicated by the consistency test result of 82%. The sludge cement is found to be quick setting, with the initial set occurring before the specified 45 min. The final setting times are well within the limiting value of 8 h.

The compressive strength results for 50-mm mortar cubes are shown in Figure 5.3. The results indicate that air-cured cement mortar exhibits the highest strength obtained at all ages. The 7-day and 28-day strength are 5.92 and 6.28 N/mm², respectively. It may be noted that ASTM C91 Standard Specification for Masonry Cement requires 3.45 N/mm² at 7 days and 6.21 N/mm² at 28 days. The strength of sludge cement under air curing is adequate for general masonry work.

The results presented in this study show that cement made from mixtures of sludge and limestone in equal amounts by weight, fired at 1000°C for 4 h under controlled firing, could be used as a masonry binder. But the effects of high water demand and quick setting of the cement must be studied. However, before the use of sludge in making masonry binder becomes commercially viable, it is necessary to carry out more research work in determining the quality of the mortar based on its air content and water retention characteristics. Further studies on long-term properties, such as durability are necessary prior to acceptance as a suitable masonry binder.

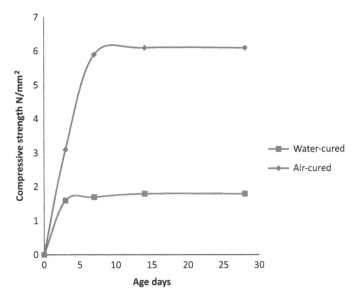

Figure 5.3.   Mortar strength

Luz *et al.* (2009) investigated the application of industrial waste obtained from the chromium electroplating process, galvanic sludge, as secondary raw materials in the production of sulfoaluminate cement. The blended sulfoaluminate cement exhibited acceptable compressive strength at around 32 MPa which has been widely used in the construction industries.

5.3 Summary

Sludge contains the basic ingredients that are needed for cement production, such as SiO_2, Al_2O_3, and Fe_2O_3. The lack of pozzolanic property in some sludge ash resulted in ineffective cement production; hence it can only be added in small amounts as inert filler in concrete or other building materials. The sludge ash samples obtained by Tay (1987b, 1987c) through pulverizing hardened sludge ash fired at 550°C are rather inert in nature; therefore it can only be incorporated in cement as filler for disposal purpose without any improvement in the compressive strength.

The results obtained from the study indicated that sludge ash could be used for up to 10% as partial replacement of cement for concrete mixing to maintain the strength of normal concrete. The addition of pulverized sludge ash filler was observed to reduce compressive strength and to some extent improve the workability of the concrete, with little effect on segregation. However, long-term properties such as durability and volume change should be studied prior to its acceptance as suitable concrete filler.

Bhatty and Reid (1989a), who obtained the sewage sludge ash in powder form directly from the incinerator for similar application, found that the sludge ash has a finely dispersed and porous nature and lacks pozzolanic property. The addition of this type of sludge ash resulted in a decrease in mortar strength and workability. It was found that the addition of fly ash reverses the detrimental effects; however it cannot provide the same strength as compared to sand mortar.

Monzo *et al.* (1996) examined the incorporation of sludge ash sample obtained in powder form directly form the incinerator and observed a negative influence on the workability of mortar mixes even thought there was an improvement in mortar strength. On further investigation, Monzo *et al.* (1999) observed possible pozzolanic properties in the sludge ash samples and showed that improvement in mortar strength increased when the amount of sludge ash and the particle fineness of sludge ash increased.

Tay and Show (1991, 1992, 1993) successfully developed cementitious material from sewage sludge by burning dried sludge with limestone and clay. The cementitious material produced can be used to produce blended cements. Show (1992) found that sludge cement produced with municipal sludge, limestone and clay, when blended with ordinary Portland cement, can accelerate the setting of mortar. A higher rate of strength development at early ages was reported by blending up to 20% by weight of sludge cement with Portland cement. With replacement of the ordinary Portland cement by up to 20% by weight, the blended cement mortar strength was not affected showing similar strength with the control Portland cement.

The study reported by Tay and Show (1993) showed that cement produced with equal portion of sludge and limestone would exhibit a compressive strength complying with the requirement of ASTM standard for masonry cement. It was found that masonry binder can be produced from dewatered sludge by mixing with limestone followed by a strictly controlled high temperature thermal treatment, producing cement that satisfies the strength requirements of the ASTM standard for masonry cement.

6

Road Construction Materials

Road pavements consist of several layers, namely subgrade, subbase, base and surface layer consisting of the wearing course and sometimes an extra binder course. Road pavement can be constructed of a variety of materials or their combinations: gravel or aggregate, stone, bitumen, concrete or improved soils. The choice of materials and thickness of the pavement layers depends mainly on the projected traffic frequency. The other factors are available budgets, road location and the availability of local materials.

Road base or base course is the main layer in sense of providing strength and load bearing capacity. Usually, road base materials are crushed and graded materials or selected natural soils. Since the function of a road base is the initial load dissipation, there are usually requirements for high-quality road base materials. In highway engineering, subbase is the layer of aggregate material laid on the subgrade, on which the base course layer is located. Subbase is often the main load-bearing layer of the pavement.

6.1 Production of Road Base Materials

A demonstration project was conducted to explore the possibility of using primary sewage sludge as road base materials in road construction (Lucena *et al.*, 2014). Addition of sewage to soil would change the characteristics of the soil. In determining the optimum amounts of sludge to be added, it is necessary to examine the mechanical behavior of the soil-sludge mixtures. Soil mixtures incorporating 10% of sewage sludge by weight producing the greatest strength property were prepared. To stabilize the soil-sludge mixtures, the effects of chemical stabilization using three conventional additives, viz. hydrated lime, Portland cement and asphalt emulsion, were examined. Four dosages (2, 4, 6, and 8% by weight) of the respective additives were investigated in order to evaluate the mechanical impact of each additive on the mixes.

6.2 Properties of Road Base Materials

The intended properties of stabilized soils depend upon the specific engineering applications. Mechanical properties of the soil, amongst others, are the most important considerations. For applications as road base materials, the soil-sludge-additive or so-called stabilized sludge-modified soils were commonly tested for the following mechanical properties: California Bearing Ratio (CBR), Unconfined Compressive Strength (UCS), Indirect Tensile Strength (ITS), Resilient Modulus (RM), and Deterioration Tests.

The CBR is used as an index of soil strength and bearing capacity, and despite the fact that it should be used for non-stabilized soils, it is a parameter broadly used and applied in the design of the base and the subbase of stabilized soils for pavements (Nicholson *et al.*, 1994). In the demonstration project, the Brazilian Standard DNIT ES-098/2007 establishes criteria, in terms of CBR values, for the use of soils in granular base layers according to the traffic (number of ESALs) that the pavement must withstand during its service life. Equivalent Single Axle Load (designated ESAL) is a concept developed from data collected at the American Association of State Highway Officials (AASHO) Road Test to establish a damage relationship for comparing the effects of axles carrying different loads.

Based on the CBR values, the soil-sludge-lime (SSL) mixture incorporating 2% lime satisfies the requirement for light traffic (ESALs $<10^6$ whereby DNIT allows the use of a base layer with a minimum CBR of 40%), and the SSL mixtures added with 4% and 6% lime meet the requirement for moderate traffic ($10^6 <$ ESALs $< 5 \times 10^6$ whereby a CBR $\geq$ 60% is required). The SSL mixture added with the highest content of lime (8%) can even satisfy the requirement for heavy traffic (ESALs $> 5 \times 10^6$ whereby a minimum CBR value of 80% is set). The results indicated that the lime-stabilized sludge-modified soils can be used for construction of asphalt base layers; the higher the content of lime, the better the soil strength and bearing capacity would be. In the soil-sludge-cement (SSC) mixtures, the stabilized SSC sample with 2% cement satisfies the requirement for low traffic, and those SSC samples added with 4% cement meet the standard for moderate traffic. CBR test was not conducted on 6% and 8% cement samples since the addition did not improve the strength of the soil. The CBR values of SSC mixtures were similar to those reported by Eren and Filiz (2009). Addition of asphalt emulsion, on the other hand, deteriorated the bearing capacity of the sludge-modified soils; the soil-sludge-emulsion (SSE) mixtures exhibited CBR values lower than the mixtures without emulsion.

Araya *et al.* (2012) reported that pavement design methodologies in developing countries are empirical in nature and rely on input parameters such as CBR, despite the fact that this parameter is not mechanistic in nature, which would allow performance prediction. As pointed out previously, CBR tests are not suitable for chemically stabilized sludge-modified soils, and empirical design methods for such mixtures typically consider compressive strength of the soil. Such a notion, however, is acceptable if the CBR results are associated with the increase in cohesion due to cementation. If soil improvement through the addition of stabilizers is associated with cationic exchange, CBR results could then be considered (Lucena *et al.*, 2014).

Unconfined compression test is used to measure the shearing resistance of cohesive soils which may be undisturbed or remolded specimens. An axial load is applied on the specimen using either strain-control or stress-control condition. The UCS is defined as the maximum unit stress obtained within the first 20% strain. The UCS results reported by Lucena *et al.* (2014) for stabilized sludge-modified soil samples with different curing methods and periods are shown in Table 6.1. The UCS tests were performed on compacted soil samples considering the optimum

Table 6.1. Unconfined compressive strength for chemically stabilized sludge-modified soils (adapted from Lucena *et al.*, 2014)

Additive (% by weight)	Unconfined compressive strength (kPa)								
	Lime stabilization			Cement stabilization			Emulsion stabilization		
	Dry cure 7-d	Wet cure 7-d	Wet cure 28-d	Dry cure 7-d	Wet cure 7-d	Wet cure 28-d	Dry cure 7-d	Wet cure 7-d	Wet cure 28-d
2	1,121	197	1,000	900	335	489	1,104	1,000	851
4	1,137	261	1,137	970	855	670	1,316	1,000	1,016
6	1,139	302	1,211	1,315	944	1,048	1,367	920	906
8	1,141	399	1,300	1,500	714	1,200	930	723	945

moisture content to achieve the dry density equivalent of Proctor of the each mixture.

Compressive strength was analyzed only until 28 days, although the intention was for it to increase over time with continued pozzolanic reactions. In general, greater strength gain was observed for wet curing of 28 days in comparing with wet curing of 7 days. It is clear that longer curing times contribute to enhanced pozzolanic reactions. Some studies considered the curing of 7 days satisfactory. Considering the strength development of the stabilized sludge-modified soils subject to dry and wet curing of 7 days, it is noted that the strength gains were not entirely due to the chemical reaction of the stabilizer. It seems that the negative pore-water pressure (soil suction) due to the presence of water in between the soil particles does cause a significant effect on pavement strength. In an investigation on soil stabilization by 2–4% of emulsion subjected to wet and dry curing, the stabilized soils with dry curing exhibited higher strength than those with wet curing for both concentrations of emulsion (Micelli *et al.*, 2007). The strength results of emulsion-stabilized sludge-modified soils shown in Table 6.1 correspond well with reported literature for emulsified granular soils (Micelli *et al.*, 2007; Santana, 2009).

The minimum compressive strength in accordance with the Illinois Department of Transportation for lime-stabilized soils in pavement base layers is set at 1.034 MPa (Lucena *et al.*, 2014). From the results presented in Table 6.1, all stabilized sludge-modified samples exceeded the required minimum strength. It has been established that a soil stabilized by lime is regarded as reactive when it gains UCS for at least 345 kPa after a curing of 28 days (Thompson, 1966). The sludge-modified soils tested in the demonstration study are reactive to lime, therefore the curing process is essential for the strength development of the stabilized soils. It should be noted that the results obtained from all the additives/stabilizers tested (hydrated lime, Portland cement and asphalt emulsion) are higher comparing to unmodified soil (113 kPa) and sludge-modified soil (156 kPa).

A study on stabilized silty soil with 5% lime indicated a lower compressive strength of 296 kPa (Araujo, 2009). In another study, the compressive strength increased with lime addition up to 4%, ranging from a minimum of 1.0 MPa and a maximum of 1.3 MPa (Farooq *et al.*, 2011).

These results are consistent with those reported in the demonstration study by Lucena *et al.* (2014). The compressive strength, however, decreased thereafter with higher lime contents. The authors attributed the strength gain and reduction by increased lime hydration and excessive lime, respectively.

Table 6.1 shows that lime-stabilized sludge-modified soils exhibit greater strength gains than the cement-stabilized counterparts. The strength values were higher than those reported with the addition of 8% cement-rice ash mix to a residual soil (Basha *et al.*, 2005), and is consistent with that reported in the study on granular soils (Santana, 2009). On the other hand, soils with a lower specific surface area required higher amounts of cement to provide greater strength and durability (Goodary *et al.*, 2012).

The ITS test is used to determine the tensile properties of the bituminous mixture which can further be related to the cracking properties of the pavement. A higher tensile strength corresponds to a stronger cracking resistance. Table 6.2 presents ITS results of stabilized sludge-modified soil samples. In examining influence of the chemical stabilization, tests were also conducted on unmodified soil with an ITS value of 30 kPa, and on sludge-modified soil with a strength of 1 kPa (Lucena *et al.*, 2014). The samples tested after 7 days of wet cure showed very low tensile resistance, and it was not possible to determine the maximum value provided by the samples. The ITS results of the sludge-modified soils without additives exhibited small soil cohesion, as expected for granular

Table 6.2.   Indirect tensile strength test for chemically stabilized sludge-modified soils (adapted from Lucena *et al.*, 2014)

Additive (% by weight)	Indirect tensile strength (kPa)					
	Lime		Cement		Emulsion	
	7-d	28-d	7-d	28-d	7-d	28-d
2	143	100	136	94	069	140
4	185	221	198	237	153	118
6	249	260	220	228	252	228
8	229	223	364	387	266	243

soils. On the other hand, the chemical stabilization increases the ITS values of the sludge-modified soils by as much as 400%.

From Table 6.2, it can be deduced that the cement is a very effective stabilizer for strength enhancement, with an optimum cement content of 8%. The ITS development of cement stabilized sludge-modified soils demonstrated the same trend as that of the UCS. For the emulsion stabilized sludge-modified soils, the ITS increases with the emulsion content, and remained constant as the emulsion increased up to 6%.

The RM is a fundamental material property used to characterize unbound pavement materials. It is a measure of material stiffness and provides a means to analyze stiffness of materials under different conditions, such as moisture, density, and stress level. It has been shown that levels of stresses correspond approximately to the conditions that the materials are subjected to at the top of the layer just below a thin coating — top of the base — under the standard axle on the pavements of low traffic volume (Marangon, 2004). In the demonstration study conducted by Lucena et al. (2014), a deviator stress of 412 kPa and a confining pressure of 137 kPa were determined. The RM increased with the content of the additives, with an optimum additive content of 8% of cement.

Loaded Wheel Test (LWT) and Wet Track Abrasion Test (WTAT) were conducted to evaluate soil deterioration upon simulated traffic (Lucena et al., 2014). Tests were performed with samples of unmodified soil, sludge-modified soil, pure sludge, and stabilized sludge-modified soils with the additive content that provided better results on mechanical tests of RCS, RM, and ITS. Soils which present high mechanical strength may be prone to cracking (Cui et al., 2014). It may occur due to hydration of the additive that induces changes in consistency and loss of the mixing water (Louw and Jones, 2015; Biswal, 2018). Furthermore, cracks may also be a product of the shrinkage of the pores caused by moisture loss during curing time (Yang et al., 2017). As the content and the additive chosen for the tests were those that provided higher strength/stiffness, it becomes necessary to conduct the deterioration tests to evaluate its susceptibility to fatigue and cracking.

All base layers of the unstabilized sludge-modified soil and cement stabilized sludge-modified soil exhibited less rut depths than the unmodified soil. The cement stabilized sludge-modified soil, though subjected to higher wheel loading, exhibited lower rut depth than all other samples. The rut depth at the 1,000th cycle of the cement stabilized sludge-modified soil was 1.8 mm, while it was 15 mm for unmodified soil base at the 20th cycle and 15.8 mm at 100th cycle for unstabilized sludge-modified soil base (Lucena *et al.*, 2014).

The wet track abrasion simulates the wearing effects of traffic and measures the amount of material that is abraded by a rotating rubber hose from a cured slurry seal pad placed under water. The amount of material lost was determined in weight of the specimen and reported as the wear value or Wet Track Abrasion Test (WTAT) loss value. Wet Track Abrasion loss for soils modified with sludge was found to be higher compared to the one with unmodified soil (Lucena *et al.*, 2014). The observed and calculated increases in weight loss in the cement stabilized sludge-modified soils are attributed to the pozzolanic activity of the cement. From the LWT and WTAT test results, it was deduced that granular sludge-modified soil stabilized by cement has greater strength behavior compared to other soils.

6.3 Summary

The review indicated that stabilization of soil modified with sludge has potential to be used in road base construction. In addition to reuse for road construction, stabilized sludge-modified soils also have promising prospect for large-scale applications involving mass construction such as ground improvement, backfilling and the construction of superstructure of buildings. Besides alleviating the problematic disposal problems, such large-scale reuse of sludge as construction materials would bring about economic, ecological and energy-saving advantages.

7

Ceramic Roof Tile Materials

Ceramic roof tiles are designed mainly to keep out rain, and are traditionally made from locally available materials such as terracotta or clay. Of the many advantages of clay roofing tile, its durability is probably the biggest one. By firing the clay roof tiles at high temperatures, this pure natural product can last for over 100 years. Another advantage is that it can be designed in a variety of shapes, sizes, and colors, adding more character and interest to the appearance of roof. It also has reflective properties such as fire-resistance and UV-resistance, which help to increase the efficiency of heating and cooling systems. This type of material is not susceptible to mold or rot, and does not shrink and expand with temperature like wood.

7.1 Ceramic Roof Tile Production

There is not just one tile-making step but a series of processes that turn the raw materials into durable construction products. The principles behind producing roof tiles are similar to bricks. A typical process flow of the tile production is shown in Figure 7.1. Raw materials are mixed and ground by milling to achieve consistency and homogeneity in particle size (Ingunza *et al.*, 2015). Water, sand, and other additives such as sludge are mixed with the clay at this stage. Ground and homogenized raw materials are pressed into custom-designed molds in which the tile products are formed. The shaped raw tiles are transported to the dryer. The drying

Finished product

Figure 7.1.   Schematic process flow of ceramic roofing tile production

process prepares the tiles for firing by extracting moisture from the soft "green" tiles to less than 2% moisture content. After drying, the "green" tiles are transferred to a furnace where they are fired in the range of 1000–1100°C. This high temperature is necessary to establish the inherent durability, strength and fire-resistance associated with the functions of the products.

The most important raw material for production of ceramic roof tiles is clay. A study was conducted to use clay from ceramic industries mixed with domestic sewage sludge for ceramic roofing tiles (Ingunza *et al.*, 2015). The sludge tested showed granulometric characteristics similar to conventional raw materials (sand and clay). For commercial roof tiles production, particle sizes of conventional raw materials normally range between 2 and 20 μm. Mineralogical composition of conventional raw materials consist predominantly of SiO_2 (75%), Al_2O_3 (15%) and K_2O

(4%). In the experiment, the dried raw sludge exhibited predominant particle sizes larger than 20 μm, due likely to higher content of organic substances. The sludge composed, on average, 71% of organic matter, 84% of solids and 1.39 of real density, with SiO_2 (32%), CaO (14%), Fe_2O_3 (13%), SO_3 (13%) and Al_2O_3 (10%) as major chemical compounds.

Based on the mineralogical compositions, it can be deduced that conventional raw material compositions were basically made up of kaolinite, quartz and feldspars (albite and microcline) mineralogical phases, while the sludge consisted mainly of quartz and feldspars (anorthite). Considering the high SO_3 content (13%) of the sludge, potential environmental risks arising from the firing process must be assessed. It has been reported that average SO_2 emissions from ceramic products incorporating sludge were below the legislative requirement (Cusido *et al.*, 2003). Nevertheless, SO_2 emissions need to be closely examined for large-scale production of ceramic roofing tile incorporating sludge.

The experimental roof tiles were produced in a commercial ceramic factory following exactly their real-scale production process, with the only difference being the partial replacement of clay by the sludge. To ensure workability and shape conformity, an average plasticity index of the mixtures was maintained at 31.3% which was within the recommended level for moulding. A total of 750 units of experimental tiles were manufactured with different concentrations of sludge (2, 4, 6, 8, and 10% of sludge dry mass) and tested.

7.2 Properties of Ceramic Roof Tiles

The experimental roof tiles were produced with full dimensions of 480 mm length, 200 mm width and 15 mm depth, which are similar to the dimensions of commercial tiles (Ingunza *et al.*, 2015). A linear relationship between the water absorption and the sludge composition was observed (Figure 7.1). Based on a best fit regression, the water absorption values of the tiles increased from 19.8% to 21.8%, corresponding to the addition of sludge from 2% to 10%. The increase in water absorption was mainly attributed to higher porosity of the finished products caused by the firing which gave rise to the release of organic matters and the formation of pores trapped within the products. Such a notion can be deduced from the fact that a high loss-on-ignition of the dried sludge (71% on average) was also recorded.

Flexural rupture strength of the tiles, however, decreased with the increase in sludge content (Figure 7.2). The flexural rupture strength of the tiles decreased from 110 to 70 kgf corresponding to the sludge content of 2% to 10%. The reduction in strength was possibly caused by higher porosity of the tile matrix, which was associated with higher organic content contributed by the sludge. Oxidation of organic matter due to the firing leads to release of gaseous products and formation of pores trapped within the matrix. A porous matrix would invariably weaken the structure

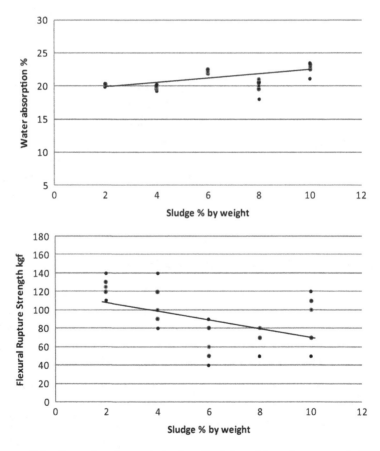

Figure 7.2. Properties of ceramic roofing tiles (adapted from Ingunza *et al.*, 2015)

of the finished products. This also explains the high water absorption of the fired products with higher sludge content. To maintain structural attribute of the finished products, flexural rupture strength may appear as a limiting factor in capping the amount of sludge in the raw mix. For acceptable strength of the tiles, the optimum content of sludge must be kept within 4% by weight of dried sludge.

The water absorption and flexural rupture strength of the tiles, however, can be improved by removing the organic matter in the dried sludge prior to mixing with clay. This can be achieved by thermal treatment of the dried sludge with temperature up to 400°C (Al-Hamati and Faris, 2014). Arising from the thermal treatment, much lower water absorption values (3.4–4.1%) were reported.

The release of gaseous products and formation of pores would be minimized when the "green" tiles are subject to subsequent firing at high temperatures. This would lead to a dense matrix and strengthened structure of the finished products.

7.3 Summary

Potential application of sludge for ceramic roofing tiles production has been demonstrated. Increasing the content of sludge would increase the water absorption but decrease the flexural rupture strength of the tiles. To maintain structural attribute of the finished products, flexural rupture strength may appear as a limiting factor in capping the amount of sludge in the raw mix. An optimum content of sludge was found to be 4% by weight of dried sludge for acceptable quality of the tiles.

Considering the high SO_3 content of the sludge, potential environmental risks arising from the firing process must be assessed. It has been reported that an average SO_2 emissions from ceramic products incorporating sludge was below the legislative requirement (Cusido *et al.*, 2003). Nevertheless, SO_2 emissions need to be closely examined for large-scale production of ceramic roofing tile incorporating sludge.

8

Leaching Characteristics of Sludge Products

Leaching is the removal of materials from the waste, usually solids such as sludge products, by dissolution, and the released constituents would then be transferred through the leachate to the surrounding environment. The effects of the constituents on living organisms, humans in particular, were determined by the concentration (Conner, 1990). Therefore, water quality standard is specified in concentration and these standards, especially drinking water standards, are usually adopted as the basis for leaching limits. Thus the rate of leaching is customarily measured and expressed in terms of concentration of the constituent in the leachant.

The amount of the constituent removed from the original waste material is of significance when evaluating the leachability of a material. The contamination potential of a material diminishes as the hazardous constituents are being leached. Thus, leaching could be a beneficial process in the long term, if the leached metal concentrations do not reach health-threatening levels (Conner, 1990).

The leaching characteristics of the toxic chemicals from the sludge products can be assessed by conducting leaching tests which simulates the field leaching conditions. The objectives of the leaching test are to evaluate the extent to which the elements concerned can be stabilized through firing and cementation (such as those cast into concrete), and to ensure that the contaminant leached out of the sludge products are within safety limits.

8.1 Leaching Test Methods

There are two common approaches in conducting leaching tests, the batch agitation method and the column-leaching method. Batch agitation method basically includes Extraction Procedure (EP) toxicity and Toxicity Characteristic Leaching Procedure (TCLP). TCLP is a method prescribed by U.S. EPA method 1311 which is basically an EP used for the testing of toxicity of a solid waste using a batch agitation concept. The TCLP test is used to determine mobility of organic and inorganic compounds present in the waste. This procedure attempts to mimic conditions a waste may be exposed to in a landfill, thus projecting the potential mobility of those compounds. It provides a sound basis for identification and disposal of non-hazardous wastes or for development of new technology that would convert hazardous wastes into non-hazardous forms.

In general, the extract from a representative sample of the waste is analyzed to determine whether it contains concentrations of one or more of the specific toxic substances exceeding the permissible limits listed by EPA. TCLP test is designed to evaluate the waste. The purpose of using acidic medium is to simulate the organic acids generated during decomposition of organic matter in such landfills. This test is particularly suitable for determining the leaching condition in static groundwater where continuous flow of leachant does not occur, as only a single elution is used and leaching decreases rapidly with time.

Methods adopting the batch agitation concept that uses a single elution do not provide reliable results. The initial elutions were believed to contain the maximum concentrations of constituents because they are exposed to the highest concentrations present on the fresh waste surface. However, this presumption may be overly conservative in the case of metal leaching (Bishop, 1988). The column-leaching test, on the other hand, generates contaminants over time. With continuous replacement of fresh leachant, the test was conducted to simulate the field leaching conditions (Francis and White, 1987; Tay *et al.*, 2001a; Show *et al.*, 2006a).

The column-leaching test enables direct concentration and the quantity of metal extracted in terms of the sample quantity to be derived for analysis. The leaching of metal is mainly governed by the leachant pH, because pH is an important factor in metal fixation which affects the concentration of metal in the leachate. Most metal hydroxides have low solubility in the range of pH 7.5–11. Some metals such as chromium, however, have higher solubility at both low and high pHs (EPA, 1987; Conner, 1990). Cement contains large amount of lime, which could raise the pH of the leachants. In the leaching of concrete samples, the pH levels were raised to the range between 10 and 12, thereby the solubility of metal hydroxides and sulfides were lowered and the leaching concentrations were reduced.

As reported by Tay *et al.* (2001a) and Show *et al.* (2006a), columns containing the test samples are leached under simulated leaching conditions in the column-leaching test. Leachate samples were prepared with distilled water to assure high dissolution of chemicals. Test samples were placed between two layers of fine sands (layers of 100 mm above and 50 mm below the sample) in perspex columns with a diameter of 150 mm and a height of 500 mm as shown in Figure 8.1. Each column contained 2.5 kg of sample except for the control column with only the sand layers.

Distilled water was pumped from each column by peristaltic pumps in a downflow mode. As the standard time of contact adopted by most protocols is 24 h (Conner, 1990), the flowrate was regulated at 1 ml/min in order to generate a total leachate volume 20 times of the sample mass, as well as to provide adequate time of contact of 30 h. In the first month of the leaching test run, leachate from the test columns was collected twice weekly; thereafter, once a week for a total of 26 samples over a period of 150 days. The pH value and electrical conductivity of the leachate were determined before preserving the samples, by adding ultra-pure concentrated nitric acid to a pH value lower than 2, for subsequent trace metal analysis. The concentrations of most elements were determined using the Ionic Coupled Plasma Atomic Emission Spectrometer. The concentrations of trace elements were analyzed using the Simultaneous Multielement Analysis Graphite Furnace Atomic Absorption system (Perkin-Elmer SIMAA 6000), and mercury was analyzed using the Flow Injection Mercury Systems.

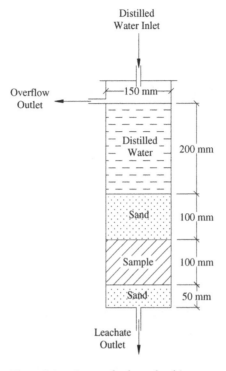

Figure 8.1.   Set up of column leaching test

8.2 Assessment of Leaching

Trauner (1993) tested the tendency of sludge ash brick samples to leach metals using the extraction procedure toxicity test (EPTox) (40 CFR 261 1988). There are mixed results on whether the use of sludge ash increased the leaching of metals as all concentrations remained well below the limits given in EPTox.

Bhatty *et al.* (1992) conducted EPA leach test on the fired sludge ash pellet aggregates found no detectable contaminants in the leachate. The pellets were reported to have a fairly neutral pH of 6.9 in an aqueous medium.

Khanbilvardi and Afshari (1995) conducted a comparison of the EP toxicity and TCLP tests to determine the characteristics of fine sludge ash aggregate samples. Results of both tests collectively showed that sludge ash is non-hazardous with most pollutants falling below the detectable limit.

Okuno *et al.* (1997) conducted a standard method for metals included in industrial waste defined by the Japanese Environmental Ministry on bricks produced from 100% sludge ash. The results showed that after treating the sludge ash at the temperature of 1020–1070°C in the process of brick making, low levels of zinc and arsenic originally present in the leachate from the sludge ash samples were diminished. No contaminants were detected in the leachate from crushed bricks, indicating that the bricks are free from environmental contamination.

Potential contamination from toxic contaminants in the concrete cast with industrial sludge aggregates using column-leaching test adopted from Francis and White (1987) was assessed (Tay *et al.*, 2001a; Show *et al.*, 2006a). The column-leaching test method was selected instead of the TCLP because the sludge products could be used for pavement and masonry purposes where the leaching behavior is most likely dynamic rather than static, such as those occurring in the landfills. Precautions were taken against inconsistent results by conducting frequent sample collection.

pH of the leachate of concrete made from sintered sludge aggregate samples were between the range of 10 and 13. The leachate tends to be alkaline due to the lime content in the cement. The leaching test was conducted for a test period of 150 days and the total mass of the full range of elements that were leached out during the test period from both the aggregate and concrete samples were calculated and presented by Hong (2000) and Show *et al.* (2006a). The levels of contamination were examined based on the health-based limits given in WHO guidelines for drinking water quality (World Health Organization, WHO, 1993). In the health-based limits, values were specified for cadmium (Cd), chromium (Cr), copper (Cu), lead (Pb), manganese (Mn), nickel (Ni) and mercury (Hg). The concentration level of cadmium was not analyzed since the element was not detected in the chemical content of the sintered aggregates.

8.2.1 *Leaching of heavy metals in sintered aggregates*

pH values of the leachate from sintered aggregates varied between 5 and 7. Peak concentrations of the contaminants leached out of sintered aggregates in pelletized- and crushed-shape were determined and are tabulated in Table 8.1. The results indicated that Cd, Cr, Cu, and Mn did not pose any form of contamination in the leachate of the sintered aggregates. This was deduced from the fact that Cd was not present at all in the aggregates and the peak concentrations of Cr, Cu, Ni, and Mn were all within the WHO limits.

A peak concentration of Hg at 0.0012 mg/L was detected in the leachate of the S/C 100–00 crushed aggregate on the fourth day of leaching. The peak concentration marginally exceeded the WHO limit of 0.001 mg/L and the concentration dropped below the limit after 7 days of leaching.

The leachate samples were detected with high levels of Pb exceeding the WHO limits for all samples including the blank. Blank samples

Table 8.1.　Peak concentrations of health-based contaminants in leachates of sintered aggregates (in mg/L)

| Element | Sludge | Blank | Aggregate sludge–clay mix (% by weight) | | | | | WHO limits |
			100–00	80–20	50–50	20–80	00–100	
(a) Pelletized aggregates								
Cr	0.03	0.01	0.04	0.03	0.03	0.03	0.02	0.05
Cu	0.06	0.01	0.02	0.02	0.02	0.03	0.03	2.00
Hg	0.0010	0.0008	0.0009	0.0007	0.0006	0.0002	0.0004	0.00010
Mn	0.09	0.01	0.07	ND	ND	ND	ND	0.50
Ni	$0.071_{(7)}$	0.004	0.004	0.005	0.006	0.006	0.007	0.02
Pb	$0.03_{(150)}$	$0.04_{(116)}$	0.04	$0.13_{(95)}$	$0.15_{(123)}$	$0.13_{(129)}$	$0.16_{(116)}$	0.01
(b) Crushed aggregates								
Cr	0.03	0.01	0.02	0.03	0.03	0.03	0.03	0.05
Cu	0.06	0.01	0.02	0.01	0.04	0.07	0.06	2.00
Hg	0.0010	0.008	$0.0012_{(7)}$	0.0008	0.0008	0.0004	0.0006	0.00010
Mn	0.09	0.01	0.09	ND	ND	ND	ND	0.50
Ni	$0.071_{(7)}$	0.004	0.004	0.005	0.007	0.008	0.008	0.02
Pb	$0.03_{(150)}$	$0.04_{(116)}$	$0.10_{(39)}$	$0.45_{(123)}$	$0.12_{(95)}$	$0.14_{(129)}$	$0.16_{(123)}$	0.01

Notes: Values in brackets are number of days taken to comply with WHO limits.
ND: Not detectable.

collected from the blank column contained only the sand layers without any specimen. The high Pb levels could be from other sources of contamination in the leaching system such as the sand or even the leaching medium itself. It could also due to the pH level of the leaching environment which ranged between 5 and 7 where solubilities of Pb hydroxide are extremely high at values above 100 mg/L (Show *et al.*, 2006a).

The peak Ni concentration detected in the leachate of sludge samples, at a value of 0.071 mg/L, exceeded the WHO limit of 0.02 mg/L. The Ni detected after 7 days of leaching would favorably diminished below the limit. The Ni concentration of the remaining leachate from sintered aggregates of various sludge–clay mixes were within the permissible limit. Hence, there appeared to be of no issue with Ni contamination that is harmful to the aquatic environment.

The Pb concentration of the blank also dropped below the WHO limit of 0.01 mg/L after 116 days of leaching. For the pelletized aggregates, leachate of the S/C 100–00 mix did not reach permissible levels within the 150 days test period, while leachate of S/C 80–20, S/C 50–50, S/C 20–80, and S/C 00–100 diminished below the WHO limit after 109, 102, 116 and 109 days, respectively. For crushed aggregates, leachate of S/C 100–00, S/C 80–20, S/C 50–50, S/C 20–80, and S/C 00–100 exceeded the limit for 139, 123, 95, 129, and 123 days, respectively, before complying with the WHO standards. It should be noted that excessive Pb in the drinking water may cause poisoning and problems leading to brain and kidney damages. Failure to comply with the limit warrant further tests in assessing suitability of the sintered sludge–clay aggregates for construction applications.

Chang *et al.* (2007) employed a sequential extraction method combined with inductively coupled plasma atomic emission spectrometry (ICP-AES) to determine the concentration and distribution of hazardous toxic elements in the metal sludge-based artificial lightweight aggregate (LWA). The results showed that the leaching concentrations of Cd, Cr, Cu, and Pb present in the non-sintered raw aggregate pellets reached 7.4, 68.0, 96.0, and 61.4 mg/L, respectively, far exceeding the regulatory threshold. Sintering at 1150°C for 15 min resulted in stronger chemical bonds formed between the elements. After the first three steps of sequential extraction, the concentrations of Cr, Cu, and Pb reached 2.69, 1.50, and 1.88 mg/L at 1150°C, while the final residues had total concentrations of 96.1, 88.4, and 60.6 mg/kg, respectively, with Cd undetected in both phases. The concentration levels fell within the regulatory threshold, indicating that the LWA fabricated from recycled metal sludge contains toxic and hazardous elements that were not leached out. Having no harmful effect on the environment, the metal sludge-based artificial LWA appeared to be safe and practical with good physical properties.

In research work on sintering behavior of dried sewage sludge (Wang *et al.*, 2008; Latosinska and Zygado, 2009; Wang *et al.*, 2009), it was reported that heavy metals are fixed primarily inside the sintered ceramic products, with metals fixed within the structure. The ash matter from sludge was encapsulated into the solid phase, with the As, Pb, Cd, Cr, Ni, Cu, and Zn concentrations in the leachate found to be in the range of local regulatory requirements. These results reveal the feasibility of recycling dried sewage sludge by sintering as a construction material.

The total mass of elements in the aggregates leached out during the 150 days test period were calculated and reported by Hong (2000) and Show *et al.* (2006a). A comparison of the total mass of contaminants leached out of the unprocessed sludge, sintered sludge and the blank is shown in Figures 8.2 and 8.3. Effect on leachability of the industrial sludge through sintering was examined and significant reductions were noted particularly on Ca, K, Mg, Na, and Si, and slight reductions were also observed on Mn and Ni. Elements of low concentration levels such as Al, Cr, Cu, Fe, Pb, and Hg did not exhibit reductions in contaminant levels due to sintering. Contaminant levels of mercury are not shown since they were of exceptionally low values.

8.2.2 Leaching of heavy metals in concrete

The leaching of concrete cast with granite aggregates (control) and sintered aggregates were evaluated. The peak concentration levels are given in Table 8.2. The leachate tended to be alkaline with pH ranged between 10 and 13. As shown in Table 8.2(a), the concentrations of most health-based contaminants diminished significantly over the leaching. As discussed earlier, the concentrations of Pb and Hg in the leachate of the aggregates which exceeded the WHO limits, but were below the limits when cast in

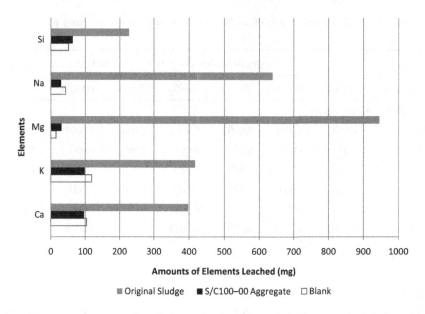

Figure 8.2.   Comparison of total major elements leached out from original sludge and S/C100–00 pelletized aggregates over 150 days of leaching

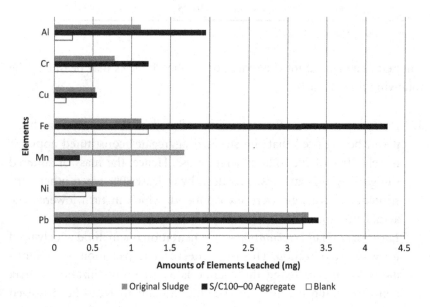

Figure 8.3.   Comparison of trace elements leached out from original sludge and S/C100–00 pelletized aggregates over 150 days of leaching

Table 8.2.   Peak concentrations of health-based contaminants in leachates of concrete (in mg/L)

Element	Blank	Aggregate sludge–clay mix (% by weight)					WHO limits
		100–00	80–20	50–50	20–80	00–100	
(a) Pelletized aggregates							
Cr	0.0001	0.0001	0.0001	0.0002	0.0002	0.0002	0.05
Cu	0.07	0.08	0.23	0.09	0.09	0.09	2.00
Hg	ND	ND	ND	ND	ND	ND	0.00010
Mn	ND	ND	ND	ND	ND	ND	0.50
Ni	0.003	0.003	0.003	0.004	0.004	0.005	0.02
Pb	0.0061	0.0040	0.0038	0.0011	0.0037	0.0010	0.01
(b) Crushed aggregates							
Cr	0.0001	0.0001	0.0001	0.0002	0.0002	0.0002	0.05
Cu	0.07	0.09	0.14	0.07	0.08	0.09	2.00
Hg	ND	ND	ND	ND	ND	ND	0.00010
Mn	ND	ND	ND	ND	ND	ND	0.50
Ni	0.003	0.003	0.003	0.005	0.005	0.006	0.02
Pb	0.0061	0.0047	0.0029	0.0053	0.0066	0.0026	0.01

Note: ND: Not detectable.

concrete. The reason for the reduction in the Pb levels may be due to the following three factors:

1. *Reduction in source of contaminant* — Based on the concrete batching, it can be deduced that the sintered aggregates constituted approximately 20% of the total concrete mass. Hence, the mass of sintered sludge–clay aggregates was reduced by at least 80% when other constituents of concrete were incorporated, which in turn lowered the actual amount of contaminants present in the source.

2. *Encapsulation by cementation* — Cementation is a method widely used for waste stabilization. This is an effective encapsulation of contaminants leading to fixing the chemicals as an immobilization matrix hence reducing the amounts of contaminant that can be dissolved through leaching.

3. *Effect of pH* — The pH level could also affect the concentration of contaminants in the leachate since leaching of metal is mainly governed by the pH. Most metal hydroxides have low solubility in the range of pH 7.5–11. Some metals such as chromium, however, have higher solubility at both low and high pHs. As cement contains large amount of lime, cementation could raise the pH of the leachants. In the concrete samples tested, the pH levels were raised to the range between 10 and 12 which in turn lowering the solubility of metal hydroxides and sulfides, resulting in the decrease of the leached metals in the leachate.

Of the toxic elements analyzed in the leachate of the concrete cast with sintered aggregates, all concentrations were within the respective limit specified by the WHO guidelines for drinking water quality. This indicates that the detrimental effects the aggregates shall have on the environment and health are insignificant. Therefore, the sintered sludge–clay products are safe for use as aggregates in concrete.

The total masses of contaminants leached from the concrete samples are shown in Figures 8.4 and 8.5. The concentrations of several elements were significantly higher in the leachate from the concrete specimens comparing to that of the aggregates alone. The high concentrations of Al, Ca, K, Na, and Si appeared to have been caused by cement and other

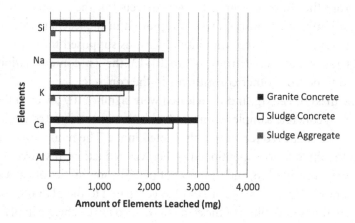

Figure 8.4. Comparison of major elements leached out from concrete samples over 150 days of leaching

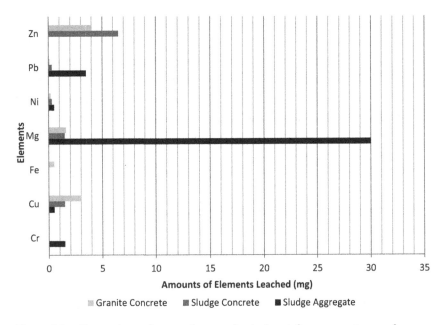

Figure 8.5. Comparison of trace elements leached out from concrete samples over 150 days of leaching

constituents of the concrete. This is because similar values were detected in the concrete control as shown in Figure 8.4.

The concentration of zinc, which was undetected earlier in the leachate from the sintered aggregates, was detected in the leachate from the concrete samples. Because Zn was also present in the leachate of the control concrete, and that the solubility of Zn was supposed to reduce when the pH range was raised from 5–7 to 10–12, the Zn contamination was deduced to be from other constituents in the concrete.

The mass of Fe, Mg, and Pb released were significantly lower in the leachate from concrete samples compared with that of the sintered aggregates alone, and a slight reduction was also noted in Cu. The reduction could be due to the three factors mentioned earlier, viz. reduction in source, encapsulation by cementation, and the effect of pH. The solubilities of Cu, Fe, and Pd are lower in the pH range of 10–12 as compared to the pH range of 5–7. The reduction in solubility of the elements caused by the change in pH could have impeding effect on the release of the contaminants (Conner, 1990).

The concentration levels of toxic chemicals in the leachate from all concrete samples complied with the limits given in the WHO guidelines for drinking water quality. The results indicate that the contamination from the sludge products is within acceptable limits and would not have any significant effects. Similar testes were also conducted on different sludge products and no potential contamination was determined.

Comparison of the concentrations of some toxic components in the sludge products determined by the studies mentioned above along with EPA guidelines and World Health Organization (WHO) drinking water limits is given in Table 8.3.

8.2.3 Leaching of heavy metals in sludge-modified soils

Peak concentrations of heavy metals leached out from the sludge and stabilized sludge-modified soil samples were determined based on leaching and solubilization tests which are tabulated in Tables 8.4 and 8.5. The heavy metal contents of sludge-modified soil samples are lower than the standard values proposed for non-polluted soils by the ABNT NBR 10004 (2004). The Mn and Ni contents in the unmodified sludge samples (Table 8.4), however, are higher than the Ministry of Health (MS, 2004) and Environmental Agency for the State of Sao Paulo (CETESB, 2005) limits for heavy metals concentrations. This implies that the unmodified sludge tested may be classified as a hazardous material. The leaching test results indicated that heavy metals did not post any form of contamination in the leachate of the stabilized sludge-modified soils. This is deduced from the fact that the heavy metals contents in all sludge-modified soil samples were lower than that of the pure sludge. The stabilized sludge-modified soils, in particular, can be classified as inert materials by the ABNT NBR 10004 (2004).

Based on the leaching test results, it is hypothesized that chemical stabilization of sludge is able to encapsulate heavy metals within the solidified matrix. Stabilized sludge-modified soils has great potential to be put to large-scale applications as base and subbase layers of pavements, ground improvement, backfilling and the construction of superstructure of buildings.

Table 8.3.   Comparison of leaching characteristic reported in various studies

Concentrations (mg/L)	Trauner (1993)	Bhatty et al. (1992)	Khanbilvardi and Afshari (1995)		Tay et al. (2001a)	WHO limit	EPA limit
Samples	Bricks	Pellets	Fine sludge ash aggregate		Concrete		
Test conducted	EP toxicity	EPA leach test	EP toxicity	TCLP	Column test		
Silver	NT	NT	<0.01	<0.01	NT	NA	<5.0
Arsenic	<0.0031	<0.05	0.16	0.12	NT	NA	<0.2/5
Barium	<0.06	<0.37	<0.2	<0.2	NT	NA	<100.0
Cadmium	<0.012	<0.01	0.06	0.16	NT	NA	<1.0
Chromium	<<0.02	<0.005	<0.01	0.01	<0.05	<0.05	<5.0
Copper	<0.097	NT	NT	NT	<0.37	<0.2	NA
Lead	<0.05	<0.01	<0.1	<0.05	<0.01	<0.01	<5.0
Mercury	NT	<0.0002	<0.0002	<0.0002	<0.0002	<0.001	<2.0
Nickel	<0.05	NT	NT	NT	<0.005	<0.02	NA
Selenium	<0.00036	<0.05	<0.19	<0.07	NT	NA	<1.0

Notes: NT: Elements not tested.
NA: Elements not available.

Table 8.4. Allowable values in the extract from the leaching test (in mg/L) (adapted from Lucena *et al.*, 2014)

Sample	Soil–sludge–cement	Unmodified sludge	Standard limits
Pb	0.20	0.33	1.000[1]
Cd	<0.05	0.05	0.500[1]
Cr	<0.10	0.10	5.000[1]
As	<0.10	0.010	1.00[1]
Cu	<0.10	<0.10	2.0[3]
Fe	<0.10	0.12	0.3[3]
Mg	—	33.20	—
Mn	0.10	0.32	0.1[2]
Ni	<0.10	0.21	0.02[3]
Zn	0.30	0.41	5[3]
Al	—	—	0.82[2]

Notes: [1]ABNT NBR 10005 (2004) and ABNT NBR 10006 (2004).
[2]Ministry of Health (MS) — Ordinance No. 518 of 2004 (MS, 2004).
[3]Environmental Sanitation Technology Company (CETESB) — Ordinance No. 195 of 2005 (CETESB, 2005).

Table 8.5. Allowable values in the extract from the solubilization test (in mg/L) (adapted from Lucena *et al.*, 2014)

Sample	Soil–sludge–cement	Unmodified sludge	Standard limits
As	<0.01	0.01	0.01[1]
Cd	<0.01	<0.05	0.05[1]
Cr	<0.001	0.1	0.005[1]
Pb	<0.01	0.39	0.01[1]
Cu	<0.10	0.10	2.0[3]
Fe	<0.10	8.52	0.3[3]
Mn	<0.10	0.77	0.1[2]
Ni	0.03	0.35	0.02[3]
Zn	<0.10	0.48	5.0[3]
Al	—	—	0.2[2]
Mg	—	43.33	—

Notes: [1]ABNT NBR 10005 (2004) and ABNT NBR 10006 (2004).
[2]Ministry of Health (MS) — Ordinance No. 518 of 2004 (MS, 2004).
[3]Environmental Sanitation Technology Company (CETESB) — Ordinance No. 195 of 2005 (CETESB, 2005).

8.3 Summary

Leaching tests for wastes are widely used in the determination of leachability of construction materials made from sludge. The values for drinking water limits are often used to ensure that the contamination from sludge product would not present any health hazard or concern. However, the testing of product leachate conducted on bricks and aggregate products is uncommon in the practice of cement-related developments. This could be due to the specific limits of chemical composition that is required in cement making that the final compound is often presumed safe. The non-leachable and non-toxic nature of the sludge products indicate that it is possible for sludge to be converted into useful products, thus eliminating contamination from sludge.

9
Challenges and Prospects

Sludge disposal is in a dynamic period of rapid change in line with the expansion of wastewater treatment plant construction and increasingly strict disposal requirements arising from industrial growth and stringent environmental regulation. With increasing population and economic growth, it can no longer rely entirely on expanding the environmental infrastructure to cope with ever-increasing amounts of sludge, nor can it entirely rely on enforcement against pollution from sludge disposal. In a number of highly urbanized cities, sludge disposal by landfilling may no longer be appropriate owing to land scarcity and increasingly stringent environmental control regulations.

Recycling nutrient-rich sludge into compost is not a quick fix either. China's Ministry of Agriculture, for example, has banned sludge-generated compost from being used on farmlands due to concerns over heavy metal contamination. And while the country's tree growers welcome the use of sludge, industry players say the market demand is limited (China Environment Forum, 2016). China's Five-Year Plan released in 2015 suggests a major shift in China's waste treatment by encouraging cities to replace landfills with more sustainable treatments (The Collective, 2017). The Five-Year Plan doesn't literally ban landfills or penalize provinces with landfills, but it does encourage "raising rates of incineration", minimizing water and gas pollution from landfills, and recycling more often. Some cities, however, have taken these pledges a step further. Beijing, Shenzhen, and Shanghai, among others, have each pledged "zero waste to landfill" by 2020 and have dedicated massive attention to waste reform.

The development of sludge management in China, the EU and other countries exemplifies global trends in viable sludge management, viz. minimization and reuse as useful resources. Incineration is one feasible means of converting bulky sludge to practically inert, odorless, and sterile ash. The disposal of this significantly reduced volume of sludge ash can be subsequently handled at a much smaller scale. The disposal problems will be drastically reduced if sludge, be it in the digested and dewatered, or the incinerated form, can be put to large-scale economic use such as construction applications. The greater use of energy- and resource-efficient technology will help minimize the adverse effect of sludge incineration. Further research on producing novel or value-added construction products from sludge by exploring the thermal methods should be considered. Instead of focusing on final treatment for sludge, pretreatment can also be considered to convert sludge into a resource carrier or construction products. It is hopeful that this review is able to serve as a reference for engineers, researchers, or policymakers in making decisions for sludge management in relation to the need for waste minimization and resource recovery.

10

Conclusions

The review of various studies on the application of sludge material in the production of building and construction materials found that it is possible to incorporate sludge into the production of construction materials while meeting the basic property requirements. Sludge material can be used in both original and ash form. The sludge ash material is usually low in plasticity and rather inert in property, while the original sludge provides a better plasticity and cohesion. It was also observed that different sludge ash may exhibit dissimilar properties. Sludge ash in the form of loose powder has a finely dispersed and porous nature while sludge ash present in the solid mass form may have a denser and impervious property.

In the production of bricks from sludge, sludge is often used in combination with clay material to achieve better plasticity and sufficient strength when dried. Studies have shown that bricks with good quality can be produced with the maximum percentages of dried municipal sludge of 40% and with the maximum percentages of municipal sludge ash of 50% by weight. Industrial sludge can be incorporated at a higher percentage of 70%. The firing temperature employed is usually within 960–1120°C. With more stringent production procedure, production of strong bricks with 100% sludge ash achieving the compressive strength of 200 N/mm^2 is possible (Okuno et al., 1997). The usual compressive strength range achieved by full-scale sludge bricks is within 18–80 N/mm^2.

Construction aggregates can also be made from sludge or sludge ash through sintering at high temperatures. Aggregates can be produced by either crushing sludge material that form a solid mass after firing under high heat or by molding and pelletizing sludge into desired shape to be directly used after firing. Clay is a popular additive to provide required plasticity for molding and pelletizing the aggregates. Aggregate made of pure sludge ash of appropriate fineness was possible (Bhatty and Reid, 1989a). The aggregate exhibited lightweight property when the material is heated to a viscous state where gases can be trapped to form cellular structure during the production.

Common firing temperatures used for producing lightweight aggregates from sewage sludge may range from 1050 to 1100°C. Higher temperature is required for materials such as clay and industrial sludge that contain lower amount of organic matter. Crushed and graded sludge ash aggregates with low thermal conductivity and high fire resistance had the required properties for production of lightweight concrete. Lightweight coarse aggregates made from pelletized or slabbed sludge ash were used in producing moderate-strength concrete. The compressive strengths that can be achieved by artificial sludge aggregates ranged from 7 to 41 N/mm^2 depending on the proportion of the aggregates and the amount of cement used.

Sludge and sludge ash may be substituted in as inert cement filler as well as a reactive pozzolanic component. Pulverized sludge ash can be added up to 10% as partial replacement of cement to maintain the strength of normal concrete. Use of non-pozzolanic sludge ash that is available in fine powdery form as filler resulted in a reduction in workability and mortar strength. For fine sludge ash that supports pozzolanic activity, improvement of mortar strength can be achieved.

Tay and Show (1991, 1992, 1993) demonstrated the production of cementitious materials from firing of sludge mix with powdered limestone. Cement made from mixtures of sludge and limestone in equal proportions by weight fired at 1000°C for 4 h exhibits the highest compressive strength under air curing condition. Evaluation of the mortar cube strength shows that it is possible to produce masonry binder made of sludge satisfying strength requirements of ASTM standard for masonry cement. The sludge masonry cement can be used to replace up to 30% by weight of ordinary Portland cement to produce blended cements without any compromise in mortar strength.

Chemical stabilization of soil modified with sludge has potential to be used in road base construction. In addition to reuse for road construction, stabilized sludge-modified soils also have promising prospect for large-scale applications involving mass construction such as ground improvement, backfilling and the construction of superstructure of buildings.

Potential application of sludge for ceramic roofing tiles production has been demonstrated. Flexural rupture strength may appear as a limiting factor in capping the amount of sludge in the raw mix. An optimum content of sludge has found to be 4% by weight of dried sludge for acceptable quality of the tiles. SO_2 emissions need to be closely examined for large-scale production of ceramic roofing tile incorporating sludge.

Leaching tests conducted on the sludge products showed no potential contamination problems for similar applications. The innovative applications of sludge as civil engineering materials effectively limit the potential sludge contamination problems and satisfy with the acceptable levels of contamination.

Converting sludge into construction materials is deemed to be a sustainable approach to alleviating sludge disposal problems and conserving natural resources. The problems of disposal and depleting raw material resources will be drastically reduced if sludge can be put to large-scale economic applications. Nevertheless, a holistic assessment of the cost involved and the benefits derived from sludge recycling has to be conducted for viable applications. The cost-benefit analysis should consider tangible and intangible factors which include energy use and costs in processing the materials, savings on raw materials replaced by sludge products, savings on land which would otherwise be used for landfilling, health hazard and environmental degradation prevented from soil and water pollution due to landfilling and natural resource extraction for raw material supply, and greenhouse gas emissions avoided from sludge recycling instead of landfilling. It is obvious that sludge applications that do not require additional processing would certainly alleviate disposal problems with economical benefits.

References

ABNT NBR 10004, 2004. *Solid Waste: Classification*, Brazilian Association of Technical Standards, Rio de Janeiro, Brazil.

ABNT NBR 10005, 2004. *Leaching of Waste: Procedure*, Brazilian Association of Technical Standards, Rio de Janeiro.

ABNT NBR 10006, 2004. *Waste Solubilization: Procedure*, Brazilian Association of Technical Standards, Rio de Janeiro.

Adams, R.B, 1988. Fate of heavy metals in sludge amended brick, *MS Thesis*, Purdue University, West Lafayette, USA.

Al-Hamati, M.F., Faris, F.G., 2014. Reuse of alum sludge in clay roof tiles manufacturing. In: *International Conference of Engineering, Information Technology, and Science* (ICEITS 2014), 17 Dec 2014, Kuala Lumpur, Malaysia. DOI: 10.13140/2.1.3713.9527.

Alleman, J.E., Berman, N.A., 1984. Constructive sludge management: Biobrick. *J Environ Eng* 110, 301–311.

Arakawa, Y., Imoto, Y., Mori, T., 1984. Cyclone furnace melting process for sewage sludge. In: *Proceedings of the 4th International Recycling Congress*, edited by Thomé-Kozmiensky, J. Karl, EF-Verlag für Energie-und Umwelttechnik, Berlin, Germany.

Araya, A.A., Huurman, M., Molenaar, A.A.A., Houben, L.J.M., 2012. Investigation of the resilient behavior of granular base materials with simple test apparatus. *Mater Struct* 45(5), 695–705.

Araujo, A.F., 2009. *Evaluation of Mixtures of Soils Stabilized with Lime, in Powder and Paste, for Application in Highways of the State of Ceara Master Dissertation*, Federal University of Ceara, Fortaleza, Brazil.

Aziz, M.A., Koe, L.C.C., 1990. Potential utilization of sewage sludge. *Water Sci Technol* 22, 277–285.

Basha, E.A., Hashim, R., Mahmud, H.B., Muntohar, A.S., 2005. Stabilization of residual soil with rice husk ash and cement. *Constr Build Mater* 19(6), 448–453.

Bhatty, J.I., Reid, K.J., 1989a. Lightweight aggregates from incinerated sludge ash. *Waste Manage Res* 7, 363–376.

Bhatty, J.I., Reid, K.J., 1989b. Compressive strength of municipal sludge ash mortars. *Mater J* 86, 394–400.

Bhatty, J.I., Malisci, A., Iwasaki, I., Reid, K.J., 1992. Sludge ash pellets as concrete aggregates in concrete. *Cement Concrete & Aggregates* 14, 55–61.

Bishop, P.L., 1988. Leaching of inorganic hazardous constituents from stabilized solidified hazardous wastes. *Hazard Waste Hazard Mater*, 5(2), 129–143.

Biswal, D.R., Sahoo, U.C., Dasm S.R., 2018. Durability and shrinkage studies of cement stabilsed granular lateritic soils, *Int J Pavement Eng*, DOI: 10.1080/10298436.2018.1433830.

Celik, T., Bayasi, Z., 1995. Performance of clay-blended sludge aggregate. *Concrete Int* 17, 63–65.

CETESB Decision of the Board No. 195, 2005. Provision for the Approval of Guiding Values for Soils and Groundwater in the State of São Paulo. DOE Executive Power SP 3/12/2005, Section 1, 115(227), 22–23.

Chang, F.C., Lo, S.L., Lee, M.Y., Ko, C.H., Lin, J.D., Huang, S.C., Wang, C.F., 2007. Leachability of metals from sludge-based artificial lightweight aggregate. *J Hazard Mater* 146, 98–105.

Chang, F.C., Lee, M. Y., Lo, S.L., Lin, J.D., 2010. Artificial aggregate made from waste stone sludge and waste silt. *J Environ Manage* 91, 2289–2294.

Chen, L., Lin, D.F., 2009. Stabilization treatment of soft subgrade soil by sewage sludge ash and cement. *J Hazard Mater* 162, 321–327.

Cherubini, F., Bargigli, S., Ulgiati, S., 2009. Life cycle assessment (LCA) of waste management strategies: Landfilling, sorting plant and incineration. *Energy* 34, 2116–2123.

Cheeseman, C.R., Virdi, G.S., 2005. Properties and microstructure of lightweight aggregate produced from sintered sewage sludge ash. *Resour Conserv Recy* 45, 18–30.

China Environment Forum, 2016. Innovative sludge-to-energy plant makes a breakthrough in China. Available on: https://www.newsecuritybeat.org/2016/05/innovative-sludge-to-energy-plant-breakthrough-china/, dated 26 Mar 2018.

Conner, J.R. 1990. *Chemical Fixation and Solidification of Hazardous Wastes.* van Nostrand Reinhold, New York, USA.

Cui, S., Blackman, B.R.K., Kinloch, A.J., Taylor, A.C., 2014. Durability of asphalt mixtures: Effect of aggregate type and adhesion promoters. *Int J Adhes Adhes* 54, 100–111.

Cusido, J.A., Cremades, L.V., González, M. 2003. Gaseous emissions from ceramics manufactured with urban sewage sludge during firing processes. *Waste Manage* 23, 273–280.

Cyr, M., Coutand, M., Clastres, P., 2007. Technology and environmental behavior of sewerage sludge ash (SSA) in cement based materials. *Cem Concr Res* 37, 1278–1289.

Department Of The Army (DOA), U.S. Army Corps of Engineers, 1984. *Engineering and Design — Soil Stabilization for Pavements — Mobilization Construction.* Engineer Manual No. 1110-3-137, Washington, D.C.

Ehlers, E.G., 1958. The mechanism of lightweight aggregate formation. *Am Ceram Soc Bull* 37, 95–99.

Elkins, B.V., Wilson, G.E., Gersberg, R.M., 1985. Complete reclamation of wastewater and sludge. *Water Sci Technol* 17, 1453–1454.

EPA Report, 1987. U.S.EPA. *Fed Reg* 52(155), 29999.

Eren, S., Filiz, M., 2009. Comparing the conventional soil stabilization methods to the consolid system used as an alternative admixture matter in Isparta Darıdere material. *Constr Build Mater* 23(7), 2473–2480.

Farooq, S.M., Rouf, M.A., Hoque, S.M.A., Ashad, S.M.A., 2011. Effect of lime and curing period on unconfined compressive strength of Gazipur soil, Bangladesh. In: *4th Annual Paper Meet and 1st Civil Engineering Congress,* Dhaka, Bangladesh.

Foley, J., Haas, D., Hartley, K., Lant, P., 2009. Comprehensive life cycle inventories of alternative wastewater treatment systems. *Water Res* 44, 1654–1666.

Francis, C.W. White, G.H., 1987. Leaching of toxic metals from incinerator ashes. *J Water Pollut Control F* 59, 979–986.

Frost and Sullivan Report, 2004. Analyst predicts sludge treatment market to grow. *Filtration Industry Analyst.*

Fytili, D., Zabaniotou, A., 2008. Utilization of sewerage sludge in EU application of old and new methods — A review. *Renewable Sustainable Energy Rev* 12, 116–140.

Garcia, R., Vigil de la Villa, R., Vegas, I., Frias, M., Sanchez de Rojas, M.I., 2008. The pozzolanic properties of paper sludge waste. *Constr Build Mater* 22, 1484–1490.

Goodary, R., Lecomte-Nana, G.L., Petit C., Smith D.S., 2012. Investigation of the strength development in cement-stabilised soils of volcanic origin. *Constr Build Mater* 28(1), 592–598.

Gonzales-Corrochano, B., Alonso-Azcarate, J., Rodas, M., 2009. Characterization of lightweight aggregates manufactured from washing aggregate sludge and fly ash. *Resour Conserv Recy* 53, 571–581.

Gray, D.H., Penessis, C., 1972. Engineering properties of sludge ash. *J Water Pollut Control Fed* 44, 847–858.

Hong, S.Y., 2000. Development of sintered sludge as construction materials, *MEng Thesis*, Nanyang Technol Univ, Singapore.

Huang, C., Pan, J.R., Liu, Y., 2005. Mixing water treatment residual with excavation waste soil in brick and artificial aggregate making. *J Environ Eng* 131, 272–277.

Ingunza, M.P.D., Pilar, M.D., Lima, A.D., Nascimento, R.M., 2015. Use of sewage sludge as raw material in the manufacture of roofs. In: *Proceedings of the 2nd International Conference on Civil, Materials and Environmental Sciences*, 13–14 Mar, 2015, London, UK. DOI:10.2991/cmes-15.2015.9.

Kato, H., Takesue, M., 1984. Manufacture of artificial fine lightweight aggregate from sewage sludge by multi-stage stream kiln. In: *Proceedings of the 4th International Recycling Congress*, edited by Thomé-Kozmiensky, J. Karl, EF-Verlag für Energie-und Umwelttechnik, Berlin, Germany.

Khanbilvardi, R., Afshari, S., 1995. Sludge ash as fine aggregate for concrete mix. *J Environ Eng* 121, 633–638.

Kosior-Kazberuk, M., 2011. Application of sewage sludge ash as partial replacement of aggregate in concrete. *Pol J Environ Stud* 20, 365–370.

Kuo, W.Y., Huang, J.S., Tan, T.E., 2007. Organo-modified reservoir sludge as fine aggregate in cement mortar. *Constr Build Mater* 21, 60–615.

Kurth, R., 1984. Sewage sludge incineration with separate drying, use of the ash and vapor disposal without chimney. In: *Proceedings of the 4th International Recycling Congress*, edited by Thomé-Kozmiensky, J. Karl, EF-Verlag für Energie-und Umwelttechnik, Berlin, Germany.

Labahn, O., 1983. *Cement Engineers' Handbook*, Bauverlag, Berlin, Germany.

Latosinska, J., Zygado, M., 2009. Effect of sewage sludge addition on porosity of lightweight expanded clay aggregate (LECA) and level of heavy metals leaching from ceramic matrix. *Environ Prot Eng* 35, 189–196.

Latosinska, J., Zygado, M., 2011. The application of sewage sludge as an expanding agent in the production of lightweight expanded clay aggregate mass. *Environ Technol* 32, 1471–1478.

Louw, S., Jones, D., 2015. Pavement recycling: Literature review on shrinkage crack mitigation in cement-stabilized pavement layers. A report prepared for California Department of Transportation Division of Research — Technical

Memorandum UCPRC-TM-2015-02, Innovation and System Information, University of California Pavement Research Center UC Davis, UC Berkeley, USA.

Lucena, L.C.F.L., Juca, J.F.T., Soares, J.B., Portela, M.G. 2014. Potential uses of sewage sludge in highway construction. *J Mater Civil Eng*, 26(9), 04014051.

Luz, C.A., Rocha, J.C., Cheriaf, M., Pera, J., 2009. Valorization of galvanic sludge in sulfoaluminate cement. *Constr Build Mater* 23, 595–601.

Marangon, M., 2004. Proposition of typical pavement structures for the region of minas gerais using local lateritic soils from pedology, MCT and resilience classification. *Doctoral Thesis*. Federal University of Rio de Janeiro, Brazil.

Metcalf and Eddy, 2004. *Wastewater Engineering: Treatment and Reuse*. 4th Edition, McGraw-Hill, International Edition, Singapore, p. 1820.

Micelli, Jr., G., Motta, L.M.G., Oliveira, J.R.M.S., Vieira, A., 2007. Resilient Behavior of Soils of the State of Rio de Janeiro Stabilized with Asphalt Emulsion. In: *21st Congress of Research and Education in Transportation*, Rio de Janeiro, Brazil.

Monzo, J., Paya, J., Borrachero, M.V., Corcoles, A., 1996. Use of sewage sludge ash (SSA) and cement admixtures in mortars. *Cem Concr Res* 26, 1389–1398.

Monzo, J., Paya, J., Borrachero, M.V., Peris-Mora, E., 1999. Mechanical behavior of mortars containing sewage sludge ash (SSA) and Portland cements with different tricalcium aluminate content. *Cem Concr Res* 29, 87–94.

Morinaga, K., Ikeno, T., Iwasaki, I., 1963. Relation between ore grindability and optimum size for pelletizing. *J Iron Steel Inst, Japan* 49, 346.

Ministry of Health (MS) Ordinance No. 518/2004, 2004. Establishes the procedures and responsibilities related to the control and surveillance of water quality for human consumption and its standard of portability, and other measures, Brazil.

Nicholson, P., Kashyap, V., Fuji, C., 1994. Lime and fly ash admixture improvement of tropical Hawaiian soils. *Transportation Research Record 1440*, National Research Council, Washington, D.C.

Norton, F.H., 1970. *Fine Ceramics, Technology and Applications*. McGraw-Hill, New York.

Okuno, N., Takahashi, S., Asada, S., 1997. Full scale application of manufacturing bricks from sewage. *Water Sci Technol* 36, 243–250.

Omer, A. M., 2008. Energy, environment and sustainable development. *Renewable Sustainable Energy Rev* 12, 2265–2300.

Rao, S.M., Reddy, B.V.V., Lakshmikanth, S., Ambika, N.S., 2009. Re-use of fluoride contaminated bone char sludge in concrete. *J Hazard Mater* 166, 751–756.

Riley, C.M., 1951. Relation of chemical properties to the bloating of clays. *J Am Ceram-Soc* 34, 121–128.

Santana, W.C., 2009. Contribution to the study of soil — emulsion in pavements of low traffic highways for the State of Maranhão. *Doctoral Thesis*, Polytechnic School of University of São Paulo, Brazil.

Show, K.Y., 1992. Properties of lime-blended sludge ash as a cementitious construction material. *MEng Thesis*, Nanyang Technol Univ, Singapore.

Show, K.Y., Tay, J.H., 2004. Conversion of sludge into novel materials for construction applications. *Trans Mater Res Soc Jpn* 29, 1957–1960.

Show, K.Y., Lee, D.J., Tay, J.H., Hong, S.Y., Chien, C.Y., 2005. Lightweight aggregates from industrial sludge-marine clay mixes. *J Environ Eng* 131, 1106–1113.

Show, K.Y., Tay, J.H., Lee, D.J., 2006(a). Properties of high strength aggregates from clay-amended industrial sludge. *J Residue Sci Technol* 3, 113–123.

Show, K.Y., Tay, J.H., Lee, D.J., Hong, S.Y., Chien, C.Y., 2006(b). Thermochemical analysis in the pre-sintering phase of aggregates from sludge-marine clay mixes. *J Mater Civil Eng* 18, 55–60.

Slim, J.A., Wakefield, R.W., 1991. The utilisation of sewage sludge in the manufacture of clay bricks. *Water South Africa* 17, 197–202.

Tay, J.H., 1987(a). Bricks manufactured from sludge. *J Environ Eng* 113, 278–283.

Tay, J.H., 1987(b). Sludge ash as filler for Portland cement concrete. *J Environ Eng* 113, 345–351.

Tay, J.H., 1987(c). Properties of pulverized sludge ash blended cement. *Mater J* 84, 358–364.

Tay, J.H., Yip, W.K., 1988. Lightweight concrete made with sludge ash. *Struct Eng Rev* 1, 49–54.

Tay, J.H., Yip, W.K., 1989. Sludge ash as lightweight concrete material. *J Environ Eng* 115, 56–64.

Tay, J.H., Yip, W.K., Show, K.Y., 1991. Clay-blended sludge as lightweight aggregate concrete material. *J Environ Eng* 117, 834–844.

Tay, J.H., Show, K.Y., 1991. Properties of cement made from sludge. *J Environ Eng* 117, 236–246.

Tay, J.H., Show, K.Y., 1992(a). The use of lime-blended sludge for production of cementitious material. *Water Env Res* 4, 6–12.

Tay, J.H., Show, K.Y., 1992(b). Utilisation of municipal wastewater sludge as building and construction materials. *Resour Conserv Recy* 6, 191–204.

Tay, J.H., Show, K.Y., 1993. Manufacture of cement from sewage sludge. *J Mater Civil Eng* 5, 19–29.

Tay, J.H., Show, K.Y., 1999. Reclamation of wastewater sludge as innovative building and construction materials. In: *4th World Congress on R'99 — Recovery, Recycling*. Reintegration, Geneva, Switzerland.

Tay, J.H., Hong, S.Y., Show, K.Y., 2000. Reuse of industrial sludge as pelletized aggregate for concrete. *J Environ Eng* 126, 279–287.

Tay, J.H., Show, K.Y., Hong, S.Y., 2001(a). Concrete aggregates made from sludge-marine clay mixes. *J Mater Civil Eng* 14, 392–398.

Tay, J.H., Show, K.Y., Hong, S.Y., 2001(b). Reuse of industrial sludge as construction aggregates. *Water Sci Technol* 44, 269–272.

Tay, J.H., Show, K.Y., Hong, S.Y., 2002(a). Concrete aggregate made from sludge-marine clay mixes. *J Mater Civil Eng* 14, 392–398.

Tay, J.H., Hong, S.Y., Show, K.Y., Chien, C.Y., Lee, D.J., 2002(b). Manufacturing lightweight aggregates from industrial sludges and marine clay with addition of sodium salt. *Water Sci Technol* 47, 173–178.

Tay, J.H., Show, K.Y., Hong, S.Y., Chien, C.Y., Lee, D.J., 2003. Thermal stabilization of iron-rich sludge for high strength aggregates. *J Mater Civil Eng* 15, 577–585.

Tay, J.H., Show, K.Y., Lee, D.J., Hong, S.Y., 2004. Reuse of wastewater sludge with marine clay as a new resource of construction aggregates. *Water Sci Technol* 50, 189–196.

The Collective, 2017. China's landfills are closing: Where will the waste go? http://www.coresponsibility.com/shanghai-landfills-closures/, dated 26 Mar 2018.

Thompson, M.R., 1966. Lime reactivity of Illinois soils. *J Soil Mech Found Div. ASCE*, 92(5), 67–92.

Trauner, E.J., 1993. Sludge ash brick fired to above and below ash vitrifying temperature. *J Environ Eng* 119, 506–519.

Utley, R.W., Lovell, H.L., Spicer, T.S., 1985. The preparation of coal refuse for the manufacture of lightweight aggregate. *Trans Soc Min Eng*, 346–352.

Wang, X., Jin, Y., Wang, Z., Mahar, R.B., Nie, Y., 2008. A research on sintering characteristics and mechanisms of dried sewage sludge. *J Hazard Mater* 160, 489–494.

Wang, X., Jin, Y., Wang, Z., Nie, Y., Huang, Q., Wang, Q., 2009. Development of lightweight aggregate from dry sewage sludge and coal ash. *Waste Manage* 29, 1330–1335.

Wang, Y., Liao, Y., 2011. Analysis quantitatively of the quality of solidified waste matrices with the ratio of binder to waste. *Adv Mater Res* 243–249, 2469–2472.

World Health Organisation WHO, 1993. *Revision of the WHO Guidelines for Drinking Water Quality*. World Health Organisation, Geneva.

Yang, J., Wang, Q., Zhou, Y., 2017. Influence of curing time on the drying shrinkage of concretes with different binders and water-to-binder ratios. *Adv Mater Sci Eng* 2017, 2695435. DOI: https://doi.org/10.1155/2017/2695435.

Yip, W.K., Tay, J.H., 1990. Aggregate made from incinerated sludge residue. *J Mater Civil Eng* 2, 84–93.

Zakaria, M., Cabrera, J.G., 1996. Performance and durability of concrete made with demolition waste and artificial fly ash-clay aggregates. *Waste Manage* 16, 151–158.

Index

Printed in the United States
By Bookmasters